**1994 Yearbook Supplement
to McGraw-Hill's
NATIONAL ELECTRICAL
CODE® HANDBOOK**

CONTRIBUTING EDITORS

Brendan A. McPartland

Facilities Engineer
Grubb & Ellis
White Plains, New York

Steven P. McPartland

Instructor, NJ State Apprenticeship Training Program
Howell, New Jersey

Jack E. Pullizzi

Assistant Facilities Engineer
AT&T Bell Labs
Holmdel, New Jersey

Robert Germinsky

Flemington, New Jersey

1994 Yearbook Supplement to McGraw-Hill's NATIONAL ELECTRICAL CODE® HANDBOOK

Joseph F. McPartland
Electrical Consultant
Tenafly, New Jersey

Brian J. McPartland
Editor, *Electrical Contractors*
(Design and Installation Update)
Tappan, New York

Nomographs provided by Special Contributor
William C. Broderick, EE, PE
River Edge, New Jersey
201-261-3564

McGraw-Hill, Inc.
New York San Francisco Washington, D.C. Auckland Bogotá
Caracas Lisbon London Madrid Mexico City Milan
Montreal New Delhi San Juan Singapore
Sydney Tokyo Toronto

1 2 3 4 5 6 7 8 9 0   DOC/DOC   9 0 9 8 7 6 5 4

ISBN 0-07-045978-9

The sponsoring editor for this book was Harold B. Crawford, the editing supervisor was Alfred Bernardi, and the production supervisor was Suzanne W. Babeuf. It was set in Century Schoolbook by McGraw-Hill's Professional Book Group composition unit.

Printed and bound by R. R. Donnelley & Sons Company.

NATIONAL ELECTRICAL CODE® is a registered trademark of National Fire Protection Association, Inc., Quincy, Massachusetts, for a triennial electrical copyrighted publication of such corporation. The term NATIONAL ELECTRICAL CODE *as used herein means the publication constituting the* NATIONAL ELECTRICAL CODE *and is used with the permission of National Fire Protection Association, Inc. This yearbook supplement does not emanate from and is not sponsored nor authorized by the National Fire Protection Association, Inc.*

 This book is printed on recycled, acid-free stock.

Contents

We Want Your Input!

Do you have a Code question or have you recently
encountered and resolved a Code question? Then we want
to hear from you.

The *1994 Yearbook Supplement to McGraw-Hill's
NATIONAL ELECTRICAL CODE® HANDBOOK* was
developed to help you resolve the many Code-related
issues that electrical professionals face on a daily basis.
You the reader can bring to this *Yearbook Supplement* all
those valuable personal experiences in working with the
Code that can be shared with your colleagues. We
therefore invite you to describe in letter form ways
in which you have resolved problems satisfying Code
requirements. We also invite you to ask questions
regarding Code interpretation that have been
troublesome. Such questions will be answered first by
phone and then printed in future *Yearbook Supplements*.

Letters are to include your full name, address,
telephone number, and affiliation, which will be withheld
on request should your letter be reprinted in the
Yearbook Supplement. Such letters become the property
of McGraw-Hill.

Joseph F. McPartland
Brian J. McPartland

About the Authors

JOSEPH F. MCPARTLAND, "Mr. Electrical Construction," is the nation's foremost expert on electrical design and construction. He is the author of 26 books on electrical design and construction.

BRIAN J. MCPARTLAND is an electrical consultant and editor of *Electrical Contractors* "Design and Installation Update," a bimonthly technical report. He has held positions in both product engineering and sales with various electrical equipment manufacturers and was chief editor at "edi" (*Electrical Design and Installation*) magazine. Both he and Joseph F. McPartland are coauthors of *McGraw-Hill's National Electrical Code® Handbook*, 21st Edition, and *McGraw-Hill's Handbook of Electrical Construction Calculations*.

Jobsite Personnel Protection: Beware of "Open Neutrals" and "Reverse Phasing"!

Secs. 110-3 and 305-6. Jobsite GFCIs—used to protect receptacles that supply cord-and-plug connected tools—must assure personnel protection under "open neutral" and "reverse phasing" (transposed circuit conductors) conditions.

For many years The National Electrical Code (NEC) and the Occupational Safety and Health Administration (OSHA) presented virtually identical requirements for personnel protection against ground-faults on construction sites. However, a change in Sec. 305-6(a) of the 1993 NEC now extends the requirement for GFCI protection to *all* receptacles used to supply cord-and-plug connected tools on construction sites—whether the receptacles are part of the "permanent wiring" or not. Although OSHA does *not* require GFCI protection for receptacles that are part of the permanent installation, it would not prohibit such protection. Therefore, both the OSHA and NEC requirements for jobsite personnel protection can be satisfied by complying with the more stringent rule of NEC Sec. 305-6(a).

It should be noted that the development and implementation of an "assured grounding conductor program" would be recognized by both OSHA and the NEC as a substitute for GFCI protection of all 125V, 15A and 20A receptacles used to supply cord-and-plug connected tools. But, practical considerations related to the costs associated with the administration and execution of an assured grounding conductor program presents a formidable challenge. In virtually all cases, the required personnel protection is provided by GFCIs.

GFCIs commonly used for personnel ground-fault protection on construction sites range from circuit breaker types protecting an entire receptacle circuit, to GFCI receptacles used either individually or at the head of a receptacle circuit to provide "feed-through" protection for downstream (loadside) non-GFCI receptacles, to cord-connected multi-outlet GFCI assemblies or individual plug-in-type GFCIs. Although all such devices listed as Class A GFCIs will satisfy the literal wording in both the NEC [Sec. 305-6(a)] and OSHA [Section 1910.304(b)(1)(i)], OSHA enforcement officials have recently indicated that devices *not* equipped with "open neutral" and "reverse phasing" protection may *not* be used as construction-site GFCIs. And, if they are used, it will be considered a violation.

Most are probably thinking, "What the heck is 'open neutral' and 'reverse phasing' protection and what does it have to do with jobsite GFCIs?"

"Open Neutral" and "Reverse Phasing"

The term "open neutral" protection applies to an operational limitation of any GFCI device. As most are aware, GFCI devices contain electronic components that provide for sensing of ground-fault current and switching of the device contact(s). However, if the neutral conductor on the line-side of a GFCI is "opened" or lifted at a panel, the unit does not have the necessary 125V for the electronic power supply, and the GFCI device (circuit breaker or receptacle) is no longer capable of sensing or switching. And that's where the danger lies. Anyone using the receptacle(s) protected by the disabled GFCI will *not* have GFCI-type protection. But because the phase conductor is still solidly connected to the supposedly protected receptacle(s), 125V-to-ground is available at the receptacle(s). And, if a faulted drill is connected to the now unprotected receptacle, the person attempting to use that drill will be exposed to precisely the type of shock or electrocution hazard the GFCIs are intended to prevent. As a result, OSHA is now requiring that jobsite GFCIs be provided with "open neutral" protection (see Figure 1).

The term "reverse phasing" (or "interchanged circuit conductors") means that the phase and neutral have been transposed on the line-side of the GFCI device. Generally, reverse phasing presents a problem only for the breaker-type GFCI, because those devices do not switch both conductors; only the conductor connected to the phase terminal is switched. If the phase and neutral have been transposed on the line-side, the device will operate, but will open the neutral, leaving the hot leg connected to the fault (see Figure 2). With the receptacle-type GFCI, both conductors are switched when a ground-fault is sensed, so reverse phasing does not present any hazard because the device will protect personnel even if the phase and neutral are connected to the wrong terminals.

What do the NEC and UL say about this matter?

The NEC does not make any reference to "open neutral" protection for construction-site GFCIs, but recreational vehicles equipped with a single 15A or 20A, 120V circuit must have a GFCI that also provides personnel protection "under conditions of an open grounded circuit conductor or interchanged circuit conductors" [NEC Sec. 551-40(c)]. For such applications in RVs, the NEC expressly requires "open neutral" protection and "reverse phase" protection. And, because the NEC does *not* specifically call for devices equipped with those additional safety

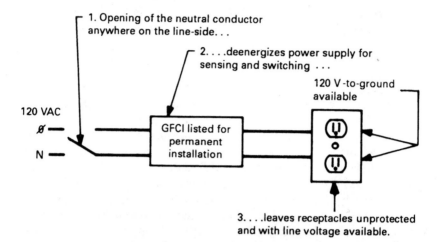

All GFCI devices listed for permanent installation
present a shock hazard under "open neutral" condition.

Figure 1 OSHA's basic concern for jobsite GFCI protection revolves around the fact that any GFCI device that is evaluated for use in a permanent installation will *not* operate when the neutral conductor is "opened" on the line-side of the device. Because the device will no longer be capable of operating, and 120V-to-ground is still available at the now unprotected receptacle, workers will unknowingly be exposed to the same type of shock hazard that the GFCIs are intended to prevent. As a result OSHA now requires that all GFCI protection for receptacles at temporary power installations be provided with "open neutral" protection.

features, it could be argued that they are *not* required. However, a look at the UL test data for GFCIs gives a different impression.

As far as UL is concerned, the need for open neutral and reverse phasing protection in any GFCI is based on the intended use. That is, if the device is intended for temporary power applications, then both open neutral and reverse phasing protection are needed for listing in accordance with UL 943. But, if the device is intended for permanent installation, other criteria apply.

The GFCI breakers listed by UL are not required to be equipped with either "open neutral" or "reverse phase" protection because where used in permanent installations, the possibility of the panel being misfed or the panelboard neutral being accidentally "opened" on the line-side is considered highly unlikely. Therefore, such features are not considered necessary for GFCI breakers, which are generally investigated and listed for use in permanent installations.

UL also generally evaluates GFCI receptacles on the basis that they will be applied in a permanent installation, unless they are submitted

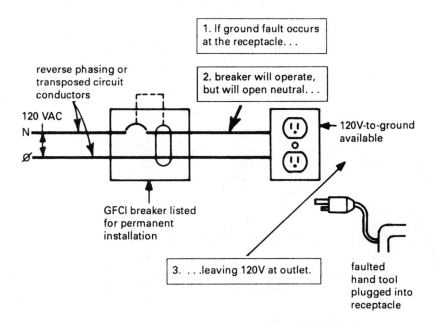

Reverse phasing of GFCI breakers listed for permanent installation represents a shock hazard because only one leg of the supply circuit is switched.

Figure 2 With GFCI-type breakers, there is also the concern for "reverse phasing" or "transposed circuit conductors" on the supply-side of the breaker because GFCI-breakers only switch the phase conductor. If the phase and neutral have been reversed on the line-side of the device, under conditions of ground-fault, the device will operate but only open the neutral conductor and leave the phase leg connected to the fault.

for temporary power use. Although an "open neutral" in a circuit supplying a GFCI receptacle would present the same problem in a permanent installation that it would with a temporary system, the chances of someone lifting the neutral at the panel in a permanent installation while someone else is attempting to use the GFCI-protected receptacle is considered extremely remote. Based on the generally accepted realities of permanent installations, UL 943 does not require "open neutral" protection for GFCI receptacles (see Figure 3).

However, because it is possible that the phase and neutral conductors could be mistakenly reversed at the receptacle's line terminals, UL does require "reverse phasing protection" for all GFCI receptacles. That is usually accomplished by switching of both conductors when a ground-fault is detected. In that way, even if the conductors were acci-

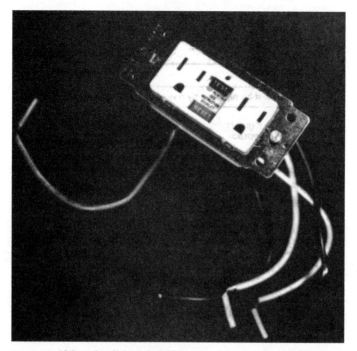

Figure 3 Although *all* listed GFCI *receptacles* are required to have "reverse phase" protection, they are not evaluated for "open neutral" protection and such an occurrence will render the device inoperative, while still supplying a lethal voltage.

dentally connected to the wrong terminals, personnel would still be protected. Therefore, all UL-listed GFCI receptacles will effectively provide "reverse phasing" protection.

As previously indicated, UL *does* require that cord-and-plug-connected GFCI assemblies—those that are manufactured and intended for use in temporary power applications—be provided with "open neutral" *and* "reverse phase" protection. These requirements effectively recognize the unusual conditions that exist in temporary wiring applications and the additional potential hazards. As a result, UL investigates these temporary-use GFCIs differently than those intended for permanent installation.

Where the cord-and-plug-connected GFCI assembly consists of GFCI-type receptacles within an enclosure, the fact that both conductors are switched by the receptacles provides the "reverse phase" protection.

Personnel protection against an "open neutral" is typically accomplished through the use of a double-pole relay. That is, the normally open contacts of a double-pole relay are connected in series with the 120V supply to the GFCI receptacles. The relay coil is also energized by the 120V supplied to the GFCI receptacles. With this wiring scheme, if the neutral is lifted at the panel or elsewhere on the supply side of the GFCI assembly, the relay coil becomes deenergized, and both the hot and neutral conductors will be disconnected from the GFCI receptacles when the relay's contacts revert to their normally open state.

It has been indicated by OSHA inspection authorities that because UL requires such additional protective features when investigating GFCI assemblies intended for temporary power applications, OSHA will now also be requiring similar protection for *all* temporary power GFCIs on jobsites. Additionally, the rule of Sec. 110-3(b) could be construed to *require* the use of listed temporary power GFCI devices and *prohibit* the use of those GFCIs listed for permanent installation on jobsites.

What to Do?

Without a doubt such interpretation of the requirements for jobsite GFCI protection by OSHA will present an additional challenge to all electrical designers and installers, but not an insurmountable one. Obviously, jobsite GFCI protection will have to be rethought, and possible solutions may prove costly. However, when compared with the dollar value of fines presently being levied by OSHA, such costs should be viewed as cheap insurance.

One solution that immediately comes to mind is the exclusive use of those GFCI assemblies that are listed and intended for temporary power applications. Because such equipment already satisfies OSHA's most recent (and rigid) interpretation, compliance can be readily achieved by using the listed cord-and-plug-connected GFCI assemblies, which are readily available.

The other approach would be the use of the "assured equipment grounding conductor program" that is given as the option to GFCI protection. Because of the administrative burden this program presents, it may not be the most desirable alternative.

Use of other GFCIs should be avoided, unless some means of compliance can be achieved by field installation, or the installation itself is otherwise arranged, for protecting personnel in the event of an "open neutral" or "reverse phasing." For example, where breaker-type GFCIs are used, it may be considered acceptable to verify that the breakers have been properly fed, and that the phase leg is being switched by testing the breaker and checking for voltage prior to use. If it can be demonstrated that "reverse phasing" has been avoided and that fact can be documented—by recording the results of each breaker's test—the OSHA inspector may consider that the requirement has been satisfied.

Where GFCI receptacles are used, a contactor wired in series with the panel feeder (from which the receptacles are supplied) that deenergizes the panel if the feeder neutral were lifted, should serve to provide protection against an "open neutral." That same concept could be applied to a panel of GFCI breakers to provide "open neutral" protection, but the concern for "reverse phasing" must still be addressed.

Even if acceptable to the local OSHA and electrical inspector, the use of "homemade" assemblies should, to the maximum extent possible, be eliminated. Such an approach will assure the minimum amount of legal exposure in the event of an accident or incident.

Although there is no doubt that such interpretation of construction-site GFCI protection goes way beyond that which is indicated by the wording of both NEC and OSHA rules covering GFCI protection for temporary wiring installations, be advised that OSHA will be enforcing this interpretation. Additionally, the concern for limiting legal exposure seems to call for the exclusive use of GFCIs specifically listed for temporary power applications. Although each of us must make our own decision regarding the handling of this matter, it is clear that the most prudent approach is to assure personnel jobsite safety even under open neutral and reverse phasing conditions.

Harmonics and the 1993 Code: The Double-Edged Sword

Secs. 210-4(a) and 220-22. Although well intentioned, the new FPN following Sec. 210-4(a) and Sec. 220-22 places designers and installers in a precarious position.

Over the past decade, most electrical designers and installers have become aware of the destructive potential that harmonic currents represent to certain distribution system components. Indeed, harmonics have been the topic of many seminars and workshops. And the electrical-trade magazines have published numerous articles as well as the results of real-world studies to provide a greater understanding of the nature and extent of the harmonic problem.

Because the concern for harmonics has grown—especially over the past three years—there were a number of proposals submitted for inclusion in the 1993 National Electrical Code (NEC). Those proposals were aimed at establishing hard-and-fast rules for dealing with the harmonic pollution that is present in virtually every commercial, industrial, and institutional electrical system in the country today. However, even though the nature and extent of the harmonic problem has been better defined and understood, there still remains a number of questions and a fair amount of disagreement related to the extent of the problem, as well as what remedies may be required to assure that any installation is "essentially free from hazard" (Sec. 90-1). And those who have studied harmonic currents and the various problems they present will be the first to say that much remains to be learned. That is true of both the need for, and the effectiveness of, the many different methods and new equipment that are being developed and used. Only through a thorough and complete unbiased evaluation in a good number of actual applications can any method or equipment be truly proven effective. But, even when the various methods or equipment are determined to be effective, there is still the question of "where" and "when" those methods or equipment *must* be used.

Given the fact that much remains to be understood, the Code Making Panels rejected proposals for new rules to address the concern for harmonics in present-day electrical systems. That action is to be commended. To incorporate *mandatory* requirements at this time is premature. That position becomes easier to understand when one considers the "derating concept" that was initially recommended when harmonics were first identified as the cause of insulation failure in transformers. Today, the concept of derating a transformer's kVA rating is not only undesirable, it can be viewed as a violation of Sec. 110-3(b). Precisely because the recommended methods and available equipment are continuing to evolve, what was the consensus recommendation yesterday may be considered objectionable or unnecessary today.

Another action taken was the establishment of a subcommittee whose directive is to study the harmonic problem and make recommendations for rule changes. This is another good move, albeit somewhat overdue. The formation of such a subcommittee will help to accelerate the journey along the learning curve by increasing awareness, providing impetus for all segments of the electrical industry to undertake studies and/or make resources available for further research, and will provide a central point for analysis and dissemination of information. Certainly, this action should have been taken some time ago. But, as the saying goes, "Better late than never."

Unfortunately, the CMPs took one further step—they added a Fine Print Note in two sections. The new FPN added after Sec. 210-4(a), "Multiwire Branch Circuits," and Sec. 220-22, "Feeder Neutrals," reads as follows:

> A 3-phase, 4-wire power system used to supply computer systems or other similar electronic loads may necessitate that the power system design allow for the possibility of high harmonic neutral currents.

Although the inclusion of that FPN was well intentioned, it basically says that "something" may need to be done, but doesn't say what or where. It places the burden of those determinations on the designer and/or installer. That FPN really does nothing to improve safety and should *not* have been accepted. That is, because most electrical designers and installers are aware of the harmonic problem, the FPN provides no additional or explanatory information. In fact, if someone is unaware of the harmonic problem, they will be further confused. And, in every case, it certainly puts the designer and/or installer under the gun. That is, the wording of the FPN seemingly leaves the designer and/or installer legally exposed because if something isn't done to determine whether or not any given installation is one of those that may require special design "to allow for...high harmonic neutral currents" and/or if the special design is not provided, a good lawyer will argue that the designer and/or installer was negligent. And even if the designer and/or installer did take some action to accommodate high harmonic neutral currents, the adequacy of any method used would still be subject to scrutiny because there is no indication as to what action is necessary or acceptable.

Some may be thinking, "Wait a minute! FPNs aren't mandatory requirements. They are only intended to provide additional information." While that may be true, I would hate to try explaining that distinction to a jury. They probably won't be too sympathetic.

Regardless, the FPN clearly indicates that a problem may exist and further states that some sort of special design may be necessary. And, in the mind of any reasonable person, that makes the designer and/or installer responsible for doing something—no one knows what, and in some cases, no one knows where.

Although the CMPs may have thought they were helping, in fact they have effectively told the designer and/or installer that they *may* need to do something without saying what or where. While it is agreed that high harmonic content can be viewed as a safety issue because harmonic currents have been known to cause insulation damage or failure, any individual safety concern should be addressed by specific requirements—such as in Note 10(c) to the Ampacity Tables or the wording of Sec. 220-22—*not* in a FPN, which, as we have discussed, is *not* mandatory. If the CMPs can not articulate a specific, clear rule, then they should not attempt to deal with the issue by the use of a FPN. Reference to a potential problem and the need for some sort of unspecified special design or installation technique in a FPN has the effect of presenting a rule without indicating how to satisfy it.

NEC Rules on Equipment Ground-Fault Protection

Secs. 215-10, 230-95, and 240-13.   The NEC is strict on ground-fault protection for equipment. Knowing when, where, and how to apply these ground-fault protection rules is a *must* for electrical contractors, engineers, and systems designers.

Properly selected overcurrent protective devices will generally provide protection for equipment against "bolted" (high-current, low-impedance) shorts and ground faults, as well as overload conditions. However, high-impedance ground-faults, such as the most commonly encountered arcing-type ground-faults, quite often do not produce enough current flow to cause the overcurrent protective device to trip.

This type of condition can cause severe damage to equipment and pose a hazard to personnel if the overcurrent device does not trip. For example, if a 1600A circuit is only loaded to, say, 1000A and a 400A arcing ground-fault occurs, the 1600A protective device will *not* operate because the total current draw (1400A) on the circuit is less than the long-time 1600A trip value, and significantly below the instantaneous trip value. However, documented cases of burn-downs and extensive equipment damage show that high-impedance, arcing ground-faults can be very dangerous and pose a substantial threat to property and personnel.

This problem was not addressed prior to the publication of the 1971 NEC. A proposal was submitted for inclusion in that edition, creating a new section that would specifically address the problem created by high-impedance faults. Thus, for the first time in the 1971 NEC, appeared Section 230-95, requiring ground-fault protection for equipment.

It is important to distinguish between "equipment ground-fault protection" and "personnel ground-fault protection." The NEC, in a number of sections, requires either *equipment* ground-fault protection or *personnel* ground-fault protection. There are primarily two points of distinction between personnel and equipment ground-fault protection.

First, each type of ground-fault protection is intended for different purposes. Personnel ground-fault protection is aimed at preventing shock hazards to people; equipment ground-fault protection is intended to protect equipment. The second distinction is related to the methods used and the operating characteristics of each type. Personnel ground-fault protection is usually provided by a GFCI circuit breaker or GFCI receptacle that opens the circuit at mA-values, while equipment ground-fault protection is generally achieved by a more complex hookup of integrating (or differential or zero-sequence) current transformers, relays, and shunt-trips on the protected disconnect, and those hookups operate at current values in excess of 1200A. Although equipment ground-fault protection does offer enhanced protection to personnel because they will protect against burn, shock, or electrocution hazards that could result during equipment burn-down, the main purpose for *equipment* ground-fault protection is the protection of property (Figure 1).

**Equipment ground fault
protection required by Sec. 230-95.**

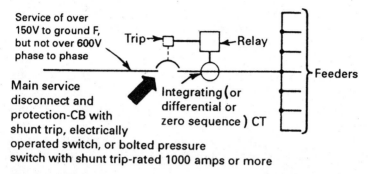

CT energizes relay to trip
disconnect on ground fault

Service of over
150V to ground F,
but not over 600V
phase to phase

Trip → □ ─ □ ← Relay

Main service
disconnect and
protection-CB with
shunt trip, electrically
operated switch, or bolted pressure
switch with shunt trip-rated 1000 amps or more

Integrating (or
differential or
zero sequence) CT

Feeders

Figure 1   This is the type of hookup required by those NEC rules calling for
equipment ground-fault protection.

What It Really Means

Essentially, Section 230-95 mandates equipment ground-fault protection "for solidly grounded wye electrical services of more than 150V to ground, but not exceeding 600V phase-to-phase for each service disconnecting means rated 1000A or more." (Figure 2)

What that means, in plain English, is that if you install or operate a solidly grounded 480Y/277V service, each service disconnect—or disconnects—rated 1000A or more, must be provided with equipment ground-fault protection.

Note that the term "rated" applies to the disconnecting means, *not* the overcurrent protection. The requirements of 230-95 must be applied if the rating of service disconnecting means exceeds, or is capable of exceeding, 1000A. Simply put, a 1200A-rated service disconnect with 900A fuses requires ground-fault protection because the rating of the *disconnect* (the fusible switch) exceeds the 1000A level at which equipment ground-fault protection becomes mandatory. Similarly, a 1000A-frame adjustable circuit breaker used as a service disconnecting means with an 800A trip setting requires ground-fault protection because the rating of the *disconnect* (the circuit breaker switching mechanism) is 1000A (Figure 3).

Ground-fault protection is *not* required for ungrounded delta, ungrounded wye, resistance-grounded wye, or red-leg delta services. Nor is it required on 480Y/277V services if the service disconnect(s) are rated less than 1000A, keeping in mind the preceding discussion on rat-

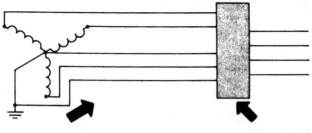

. . . applies to solidly grounded wye services over 150 volts to ground but not over 600 volts— i.e., 480Y/277 volts.

For *each* service disconnect rated 1000 amps or more, ground-fault protection with maximum trip setting of 1200 amps must be provided.

GFP IS NOT MANDATORY FOR

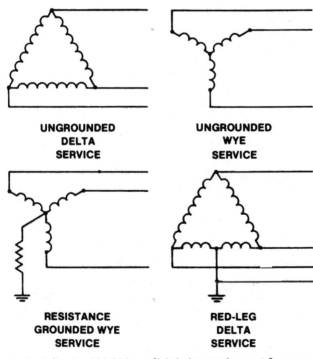

UNGROUNDED
DELTA
SERVICE

UNGROUNDED
WYE
SERVICE

RESISTANCE
GROUNDED WYE
SERVICE

RED-LEG
DELTA
SERVICE

Figure 2 Section 230-95 is explicit in its requirement for ground-fault protection of equipment. As can be seen, only 480Y/277V services with disconnects rated 1000A or more must have this protection.

FUSED SWITCH (bolted pressure switch, service protector, etc.)

Rating of switch is taken as...   ...the amp rating of the largest
fuse that can be installed in the
switch fuseholders.

EXAMPLE

If 900-amp fuses are used in this service switch, ground-fault protection would be required, because the switch can take fuses rated 1200 amps—which is above the 1000-amp level at which GFP becomes mandatory.

CIRCUIT BREAKER

Rating of breaker   ... the maximum continuous current rating
is taken as...   (pickup of long time-delay) for which the
                          trip device in the breaker is set or can
                          be adjusted.

Example: GFP would be required for a service CB with, say, an 800-amp trip setting if the CB had a trip device that can be adjusted to 1000 amps or more.

Figure 3 There is a clear method for determining the need for ground-fault protection. The rating of the fused switch or circuit breaker must be applied, rather than the actual rating of fuse or setting on a CB. If the circuit breaker or fused switch is capable of being set or is rated at 1000A or more, ground-fault protection is required.

ings. Ground-fault protection is *not* required on 208Y/120V services nor is it required on those services operating at over 600V phase-to-phase.

It should be noted that design of the service can play a part in determining the need for ground-fault protection. For instance, a 1200A-rated 480Y/277V service with a single, 1200A disconnect requires ground-fault protection per Section 230-95. That same 1200A-rated service using three 400A disconnects would not require ground-fault protection, because the service disconnecting means are rated less than 1000A (Figure 4).

There are two Exceptions to 230-95. Exception No. 1 eliminates ground-fault protection on a service disconnecting means for a continuous industrial process where a sudden shutdown (nonorderly) would

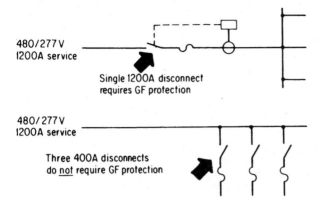

480/277V
1200A service

Single 1200A disconnect
requires GF protection

480/277V
1200A service

Three 400A disconnects
do <u>not</u> require GF protection

Figure 4 This engineering option illustrates how ground-fault protection would not be required on this 1200A service, if multiple lower-rated service disconnects were installed, instead of a single 1200A main disconnect.

create or increase hazards to the equipment. Exception No. 2 eliminates the need for ground-fault protection on fire pumps.

The reasoning behind Exception No. 1 is based on the concept that locations using continuous processes usually have highly trained, qualified personnel on-site. Maintenance and design are such that electrical safety goals can be readily achieved without the use of automatic ground-fault protection.

In addition, experience has shown that suddenly shutting down a continuous process can create more problems than it solves. There is the potential hazard to workers, as well as the possibility of machine damage, which could also increase the chances of an accident.

Fire pumps must be able to run at all times. They are required to have overcurrent protection large enough to permit running into locked rotor conditions in order to perform their task. This is per NEC Sec. 230-90, Exception No. 4. This concern is also noted in Sec. 430-31, "Part C. Motor and Branch-Circuit Overload Protection." The last paragraph of that section notes that "These provisions shall not be interpreted as requiring overload protection where it might introduce additional or increased hazards, *as in the case of fire pumps*" (emphasis added). The logic here is that it is better to let the motor burn out while doing, or attempting to do, its job, rather than trying to "protect" it from destruction. If the building is on fire, what good is saving the fire pump?

The waiving of ground-fault protection for fire pump service disconnect is born from the same logic that eliminates the need for overload protection of fire pumps. Because many larger-size pumps (i.e., those over 100 hp) would have service disconnects rated 1000A or more, they

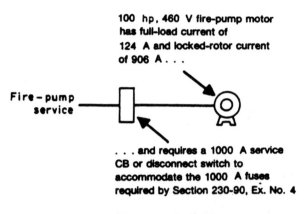

100 hp, 460 V fire-pump motor has full-load current of 124 A and locked-rotor current of 906 A . . .

Fire - pump
service

. . . and requires a 1000 A service CB or disconnect switch to accommodate the 1000 A fuses required by Section 230-90, Ex. No. 4

GROUND-FAULT PROTECTION MUST NOT BE USED!

Figure 5   Fire pump service disconnect is *not* required to be provided with equipment GFP (Sec. 250-95, Ex. No. 2).

would normally be subject to the rules of Sec. 230-95. But, Ex. No. 2 to Sec. 230-95 eliminates all requirements for equipment ground-fault protection of fire pump service disconnects (Figure 5).

Evolution

Like many sections of the NEC, 230-95 has evolved over time, being refined, updated, and amended so that it meets and addresses the current state of technology as well as the practical needs of those in the field. As originally written, for example, Section 230-95 contained no Exceptions or Fine Print Notes.

Exception No. 1 for a continuous industrial process first appeared in the 1978 NEC. Ex. No. 2 made its debut in the 1981 edition. Interestingly, the requirements spelling out the maximum setting, time-delay, and maximum values of ground-fault currents were not incorporated into the 1971 Code.

These requirements, now found in subsection (a), mandate that the selected protection be set to operate when the ground-fault current reaches, or exceeds, 1200A and must open all of the ungrounded conductors. It also specifies a maximum time delay of one second for ground-fault currents equal to or in excess of 3000A. These specific operational requirements did not appear until the 1978 version of the NEC was published.

Subsection (c), describing performance testing, was new for the 1975 NEC. This subsection requires that a performance test be performed following installation. This clearly means the test must be performed on the job-site. Factory testing, although well and good, will not satisfy the requirements of this portion of the Code. The test is to be conducted in accordance with instructions that must be provided with the equipment. Additionally, a written record of the test must be made, and this test record must be made available to the authority having jurisdiction.

The FPNs

Sec. 230-95 contains six Fine Print Notes. Without going into the history of each one, the FPNs define some terms as used in 230-95, and also provide some advisories concerning the use of ground-fault protection on other types of systems. Additionally, the FPNs make an attempt to clarify the intent of the Code Making Panel (CMP), when they formulated this rule.

For example, FPN No. 2 says that it may be desirable to provide ground-fault protection for a service disconnecting means rated less than 1000A on a solidly grounded wye system operating at more than 150V to ground, but not over 600V phase-to-phase. Basically, this FPN is encouraging the use of ground-fault protection on *any* service disconnecting means installed on a 480Y/277V service. However, the FPN is advisory only. It does not have the force of law. In other words, an inspector *cannot require* such ground-fault protection on services where the disconnecting means are rated less than 1000A. So while the CMP says it would be nice, and would indeed provide extra protection, it is not required.

One of the more significant FPNs is No. 4. This statement points out that no protection is provided on the line-side (supply side) of the ground-fault sensing CTs. That is, if a fault occurs somewhere within the equipment ahead of the sensing CTs, the ground-fault hookup will *not* open the service disconnect. Because the wording of the basic rule requires protection for "each service disconnecting means," it seems that compliance can only be assured by locating the sensing CTs ahead of the service disconnect. That way even if a fault occurs within the service disconnect, the GFP protective hookup will attempt to operate and open the service disconnect.

FPN No. 6 discusses those installations where ground-fault protection is provided for a service disconnecting means as required by 230-95, and there exists an interconnection with another supply system through a transfer switch. The Note says that means or devices may be needed to assure proper ground-fault sensing by the ground-fault protection equipment. This note is intended to communicate the idea that on-site generators, or other sources of supply—that are grounded and bonded as separately derived systems in accordance with Sec. 250-26—need to have 4-pole transfer switches to prevent desensitizing of the equipment ground-fault protection (Figure 6).

In Art. 700, the NEC modifies equipment ground-fault protection for generators and other sources of supply that are considered to be "Emergency Systems." Sec. 700-7, titled "Signals," requires that "audible and visual signal devices shall be provided, where practicable for the following purposes..." Among those purposes is subsection (d), "Ground Fault."

Sec. 700-7(d) requires means of providing an *indication* that a ground-fault exists in solidly grounded wye emergency systems of more than 150V to ground and circuit protective devices rated 1000A or more. Note, however, that ground-fault *protection* is not required for emergency systems, meaning automatic interruption of the fault is *not* required. This is made clear in Section 700-26, which states that the alternate source for emergency systems shall not be required to have ground-fault protection for equipment. Only an indication of a ground-fault is needed for emergency source disconnects. Further, Sec. 700-7(d) *does* specifically require that the sensor for the ground-fault signal devices be located at, or ahead of, the main system disconnecting means for the emergency source (Figure 7).

What is most interesting about the wording of the first sentence in Sec. 700-7 is the use of the word "practicable." This seems to indicate that if it were not "practicable" to install such signalling devices, then ground-fault indication for an "Emergency System"—generator or other source of supply—is not required. The term is not defined within the context of the requirement, and obviously, such determination would no doubt be up to the authority having jurisdiction.

A similar concept is applied to "Legally Required Standby Systems" as covered in Art. 701. Again, the Code only calls for ground-fault indication. And Sec. 701-17 uses the same wording found in Sec. 700-26 to indicate that the source disconnect of a "Legally Required Standby System" is *not* required to be provided with equipment ground-fault protection.

Art. 702, which covers "Optional Standby Systems," makes no mention of equipment ground-fault protection for those on-site generators or other power sources that would be considered as "optional" power systems. But, as covered in Sec. 90-3, all the rules of Chapters 1 through 4 of the NEC apply to those installations covered in Chapter 7, *unless* the general rules are specifically amended or modified "for the particular conditions." Art. 702 does not modify or amend the general requirements for equipment ground-fault protection. Therefore, for "Optional Standby Systems" (i.e., those covered by Art. 702), the rules for equipment ground-fault protection *must be* satisfied in all respects.

3. This GF current coming back on neutral goes through GFP sensor and is not sensed as fault current.

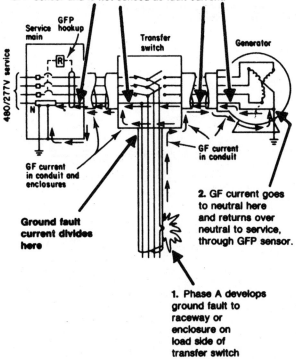

Improper operation of GFP equipment can result from emergency system transfer switch.

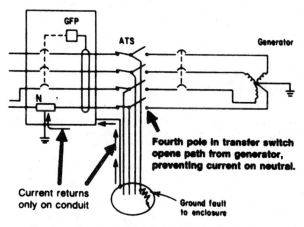

Four-pole transfer switch eliminates desensitizing of equipment GFP.

Figure 6 Where alternate means of supply are installed, desensitizing of ground-fault sensing can occur.

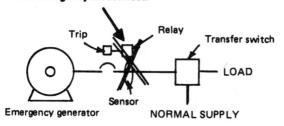

**Ground-fault protection is *not* required
for emergency disconnect.**

. . . But, Sec. 700-7 (d) REQUIRES ALARM ON FAULT.

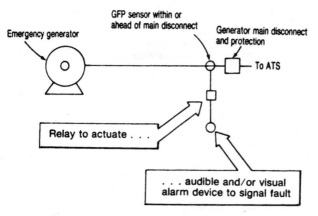

Figure 7 For emergency power systems, an alarm indication is
required instead of automatic interruption.

Additional Requirements

To further address the need for equipment ground-fault protection on
electrical systems, two new sections were added to the 1990 NEC.
Specifically, these are Secs. 215-10 and 240-13, which cover feeder and
main building disconnects, respectively.

Since 1971, Section 230-95 has addressed the requirement for
ground-fault protection on *service* disconnects rated 1000A or more and
operating at 480Y/277V. However, there are many installations where
feeders have disconnects and switches rated 1000A or more operating
at 480Y/277V, and the service voltage is different. As a result, those
disconnects and switches have no equipment ground-fault protection.
Prior to the 1990 NEC, there were no specific requirements for ground-
fault protection for other than *service* disconnects, but now high-capacity feeder and main building disconnects must also be provided with
equipment ground-fault protection.

Such a situation could exist if the service were medium-voltage (4160V or 13.2 kV). There would be no ground-fault protection required for that service equipment because those types of services operate at over 600V phase-to-phase, and ground-fault protection is not required under the provisions of 230-95. Similarly, on a 208Y/120V service, there would be no ground-fault protection required because the service operates at less than 150 volts to ground.

However, transformers are typically used to step the medium voltage down and create 3-phase, 4-wire, 480Y/277V feeders for lighting or other equipment installed within the structure. The same scenario would hold true for the 208Y/120V service if a transformer were supplied to step the voltage up should equipment or lighting needs dictate the use of 480Y/277V distribution.

The substantiation that accompanied the proposal for the changes in the 1990 NEC noted that ground-fault protection for services is recognized and required for systems operating at 480Y/277V, and that the new section would require the same protection for feeders operating under the same parameters, even when the service itself is not 480Y/277V. Past proposals attempted to treat these feeders as services, but as noted in the commentary, this created other concerns that could not be fully addressed by the proposed wording.

Sec. 215-10 now addresses this concern by requiring ground-fault protection for *every* feeder disconnect switch on a 480Y/277V, 3-phase, 4-wire feeder where the disconnect is rated 1000A or more. The Exception to this section states that ground-fault protection for equipment is not required if such protection is provided on the supply (line) side of the feeder. Although not clearly spelled out in the Exception to Sec. 215-10, the ground-fault protection "provided on the supply side" must effectively provide GFP for the feeder disconnect. That is, if the feeder is supplied from a transformer—either stepping voltage up or down—and ground-fault protection is voluntarily provided on the primary side of the transformer, it will *not* protect the feeder disconnect against ground-faults on the transformer secondary. Therefore, in such cases it is intended that an additional ground-fault hookup be provided for each such feeder disconnect. But, if the supply side GFP also protects the feeder disconnect against ground faults, then the feeder disconnect can be considered as adequately protected in accordance with the Exception to Sec. 215-10.

Article 240, Overcurrent Protection, also addresses the need for equipment ground-fault protection in Section 240-13, which was new for the 1990 NEC. That section also requires ground-fault protection for each disconnect rated 1000A, or more, operating at 480Y/277V that serves as a main disconnect for a building or structure.

The intent of this section is to call for ground-fault protection for a main building disconnect whether or not that disconnect is actually a "service" disconnect. This rule covers buildings supplied by a service located in another building, such as might be found in multibuilding commercial, industrial, or institutional installations under single management, or buildings supplied by a service located outdoors. Because a main building disconnect effectively acts as a "service" disconnect, ground-fault protection for that equipment must be provided (Figure 8).

Sec. 240-13 clearly states that such main building disconnects shall have ground-fault protection in accordance with Sec. 230-95. In essence, this requirement makes such main disconnects for buildings and structures no different than a service disconnect, when viewed in the context of ground-fault protection.

There are three Exceptions to 240-13. Exception No. 1 waives the requirements for ground-fault protection to a disconnecting means for a continuous process where a nonorderly shutdown will introduce additional hazards or increased risks. And, Ex. No. 2 states that the requirements of 240-13 shall not apply to fire pumps.

240-13, Ex. No. 3 eliminates the need for GFP of main building disconnects rated 1000A or more, and supplied from a 480Y/277V system if the disconnect is protected by "the service ground-fault protection" and the effectiveness of that protection is not defeated by additional connections between the grounded and grounding conductors at the outbuilding or structure, which is the basic requirement of Sec. 250-24(a). Simply put, if the building is supplied as covered in the basic rule of Sec. 250-24(a), the service ground-fault protection would be "nullified" because the neutral and ground would be bonded together at the outbuilding. That would allow any fault current within the outbuilding to get back to the neutral and *not* be "seen" as fault current by the sensing CTs in the main building. But, if the feeder neutral to the outbuilding is *not* bonded to the enclosure and an equipment grounding conductor is run with the circuit and bonded to the grounding electrode system—as is permitted by Ex. No. 2 to Sec. 250-24(a)—then the effectiveness of the "service ground-fault protection" would not be "nullified" and it would *not* be necessary to provide an additional GFP hookup for the outbuilding main disconnect.

One last point needs to be made regarding the wording of 240-13, Ex. No. 3. Notice that the literal wording eliminates the need for a main building disconnect equipment ground-fault protection where line-side GFP is provided by the "service ground-fault protection." What if there is no "service ground-fault protection," but "feeder" ground-fault protection is provided as required by Sec. 215-10? Do I need to provide an additional GFP hookup at the outbuilding if I've satisfied Ex. No. 3 in

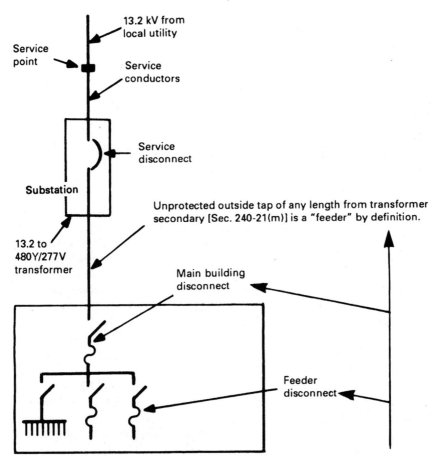

13.2 kV from local utility

Service point

Service conductors

Service disconnect

Substation

13.2 to 480Y/277V transformer

Unprotected outside tap of any length from transformer secondary [Sec. 240-21(m)] is a "feeder" by definition.

Main building disconnect

Feeder disconnect

If any of these are rated at 1000A, or more, GFP is required, unless line-side GFP protects the disconnect.

Secs. 215-10 and 240-13, which were new in the 1990 NEC, apply to feeder disconnects and main building disconnects, respectively, where they are rated at 1000A, or more, and operate on solidly grounded 480Y/277V systems.

Figure 8 Section 240-13 requires ground-fault protection for building disconnects rated 1000A or more, if they are fed from another building or from an outdoor service that is 480Y/277V. And Sec. 215-10 also calls for GFP of feeder disconnects rated 1000A or more in a 480Y/277V system. The protection must be as specified by Section 230-95.

all other respects? The logical answer should be no. Inasmuch as the same level and type of protection would be provided by either "service" or "feeder" GFP—as long as the feeder GFP is also not "nullified"—there should be no need for an additional GFP hookup even though the literal wording of 240-13, Ex. No. 3 only references "service ground-fault protection." It would certainly be better if that exception referred to "service-type" ground-fault protection or used some other such wording to clarify that "feeder" GFP, as described in Sec. 215-10, should also be acceptable provided the outbuilding is grounded and bonded as covered in Sec. 250-24(a), Ex. No. 2.

Summary

Because high-impedance faults may not generally provide enough current to trip the circuit protective device, ground-fault protection is required to protect equipment and personnel from shock and burndown caused by arcing. NEC Secs. 230-95, 215-10, and 240-13 address this concern and indicate where equipment GFP is mandatory. But, depending on the nature of the installation and its need for continuity of operation one may wish to consider the use of such equipment protection for other applications even though not specifically required.

This is especially true when emergency power systems are installed at the site. These systems, and how they may affect ground-fault sensors, has been addressed in Section 700-7(d) and 230-95 Fine Print Note No. 6. Particular care must be taken when providing ground-fault protection for equipment as required by the NEC, and an emergency power system is also being installed. The ground-fault sensing on the service must not be desensitized or compromised by emergency power systems, particularly the transfer switch.

As can be seen, the NEC requirements for ground-fault protection for equipment have evolved over a period of time, being refined and added to the Code as a result of field experience and other considerations relating to personnel safety and electrical equipment protection.

Physical Protection
for Type MC or MI
As It Emerges
from Underground?

Sec. 230-50(b), Exception and 300-5(d). The rule of Sec. 230-50(b), Ex. and Sec. 300-5(d) appear to present contradictory requirements for Type MC and Type MI cable.

Sec. 300-5(d) of the NEC requires that directly buried "cables" be provided with physical protection where they emerge from an underground trench up a pole or a building. This protection must be provided from the minimum burial depth or 18 inches, whichever is less, and must extend for no less than 8 feet above grade as required by Sec. 300-5(d). And, if the installation is subject to physical damage—such as adjacent to a driveway—the protective sleeving must be rigid metal conduit, intermediate metal conduit, or Schedule 80 nonmetallic conduit, or the equivalent (Figure 1). For installations not subject to physical damage, the cable may be protected by thinner-walled raceways, such as EMT or Schedule 40 PVC.

Although the wording of this rule clearly refers to conductors and *cables,* it is not clear as to whether this rule was intended to require supplemental protection for directly buried armored-sheathed cables—such as Type MC or MI cable—where they are not subject to physical damage. The very broad and general understanding of the term "cable" would seem to include *all* types of cable. But, when one examines the Exception to Sec. 230-50(b), it certainly seems that, where not subject to physical damage, Type MC and MI should be permitted to emerge from a trench and run up the side of a building or pole *without* additional physical protection.

The basic rule of Sec. 230-50(b) prohibits the installation of "conductors and cables other than service-entrance cables" less than 10 feet aboveground or where they would be subject to physical damage. The Exception to that rule specifically recognizes the installation of Type MC or Type MI cables "within 10 feet of grade" provided they are *not* subject to physical damage *or* where they are protected in accordance with Sec. 300-5(d). That is, Type MC and MI cables may be used to supply a service from an underground trench *without* supplemental protection, as long as it is not subject to physical damage. And, if the Type MC or MI cable *is* subject to physical damage, then they would have to be provided with rigid-steel, intermediate metal, Schedule 80 PVC, or an equivalent sleeving as described in Sec. 300-5(d).

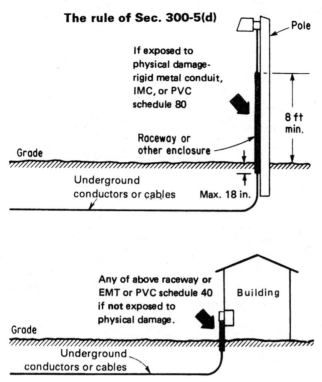

The rule of Sec. 300-5(d)

Pole

If exposed to
physical damage-
rigid metal conduit,
IMC, or PVC
schedule 80

8 ft
min.

Raceway or
other enclosure

Grade

Underground
conductors or cables   Max. 18 in.

Any of above raceway or
EMT or PVC schedule 40   Building
if **not** exposed to
physical damage.

Grade

Underground
conductors or cables

Figure 1   The rule of Sec. 230-50(b), Ex. permits "Type MC and
Type MI *cable*" to be installed in a trench and emerge from underground *without* sleeving provided the cable is not exposed to physical damage. *But,* as can be seen, the rule of Sec. 300-5(d) requires
supplemental protection for direct buried "cables" as they emerge
from underground regardless of whether or not they are exposed
to physical damage. This seems somewhat strange because the
combination of rules places greater requirements on branch circuits than service feeders. Is that what was intended?!

Part of the problem with understanding exactly what was intended
when the term "cable" was added to Sec. 300-5(d) in the 1987 NEC is
that no indication is given in available documentation as what type(s)
of cable were to be included. But, a look at some related rules seems to
justify doubt about the applicability of Sec. 300-5(d), at least for Type
MC and MI.

Consider the fact that the Type MC cable, for example, generally may
never be used where exposed to physical damage (Sec. 334-3, Uses
Permitted), and, where *not* exposed to damage, Type MC cable needs
no supplemental protection. Additionally, Type MC may be used to

supply a *service* from an underground trench *without* supplemental protection when not exposed to damage. It certainly makes little sense to require greater protection for a branch circuit to an outdoor light on a pole than for a service. Because the Code specifically requires additional protection for directly buried Type MC only in a service installation where subject to physical damage, it seems logical to conclude that similar treatment should be permitted for other applications.

This is one of many contradictions that exist in the NEC. And even though it is not possible to establish a definitive answer, such discussion serves to illuminate this and similar inconsistencies and promote discussion and analysis by many members of the electrical industry. Hopefully, through such discussion and analysis, we can arrive at some consensus of understanding and application.

Until then, be aware that the concept put forth in the preceding discussion is not clearly permitted except for services because the term "cables" is used in the wording of Sec. 300-5(d). And it will obviously be necessary to consult the local electrical inspector for an interpretation. Under the provisions of Sec. 90-4, ultimate determination of what is and what is not acceptable rests with the authority having jurisdiction. But, because the various rules are clearly inconsistent, the inspector may see fit to permit application as described here.

Conductor Ampacity and Overcurrent Protection

Secs. 240-3 and 310-15. A basic discussion of NEC rules for determining conductor ampacity and the required rating of overcurrent protection in accordance with Secs. 240-3 and 310-15 of the 1993 NEC.

One of the most fundamental tasks associated with electrical work is determining the ampacity—the Code-recognized maximum current-carrying capacity—of every circuit conductor. And that value must be established as described in NEC Sec. 310-15. Since the 1984 NEC, this procedure has been revised in each subsequent edition of the Code, and the 1993 NEC is no different, as Note 8 to the Ampacity Tables (Tables 310-16 to 310-19) has been revised.

Basically stated, Note 8 has been changed to eliminate all references to "diversity" and the derating percentages for "number of conductors" is the same as those given in "Column B" of the Table to Note 8 in the 1990 Code. The net result of all of the changes in the 1987, 1990, and 1993 NECs has been a return to the clear, logical procedures that were required by the 1984 Code, with a slight decrease in the derating percentages given in the Table to Note 8. Although it is also permissible to calculate ampacity in accordance with the Neher-McGrath method described in Sec. 310-15(b), the following discussion will focus on the reliable, straightforward, and simple procedures described by Sec. 310-15(a).

As indicated by the definition of "ampacity" given in Art. 100, the ampacity of a given conductor is that amount of current, in amperes, that the conductor can carry continuously—that is, for three hours or more without deenergizing—under specified conditions of use. The NEC approach to establishing the ampacity of any given conductor is aimed at designating the value of current that will cause the conductor to reach thermal equilibrium and stabilize at a temperature no greater than the thermal limit of the insulation.

As indicated by the wording of Sec. 310-15(a), the procedures used to establish a conductor's ampacity are based on the use of Ampacity Tables 310-16 through 310-19 and their accompanying Notes and tables. Tables 310-17 and 310-19 apply to spaced open conductors, Table 310-18 applies to high-temperature conductors, such as fixture wire, and Table 310-16 covers all other conductors—i.e., those rated 60, 75, and 90°C in raceway, cable, or directly buried in the earth. As a result, unless the conductors in question are spaced open conductors or high-temperature (over 90°C) conductors, Table 310-16 will be the table to use.

When all rules related to the procedure covered by Sec. 310-15(a) are correlated, it can be clearly established that any conductor must have its ampacity determined in accordance with one of four possible procedures.

Procedure 1

When there are *not* more than three current-carrying conductors within a conduit, cable, or directly buried, and when the ambient temperature is not over 30°C (86°F)—which are the conditions spelled out at the top of Table 310-16—then the conductor ampacity is equal to the value, in amperes, given in the appropriate Ampacity Table for the given AWG-size and insulation type. As indicated above, for the vast majority of applications using conductors rated up to 2000V, the appropriate table is Table 310-16 (Figure 1). Therefore, under the above-

Table 310-16. Allowable Ampacities of Insulated Conductors Rated 0-2000 Volts, 60° to 90°C (140° to 194°F) Not More Than Three Conductors in Raceway or Cable or Earth (Directly Buried), Based on Ambient Temperature of 30°C (86°F)

Size	Temperature Rating of Conductor. See Table 310-13.						Size
	60°C (140°F)	75°C (167°F)	90°C (194°F)	60°C (140°F)	75°C (167°F)	90°C (194°F)	
AWG kcmil	TYPES TW†, UF†	TYPES FEPW†, RH†, RHW†, THHW†, THW†, THWN†, XHHW† USE†, ZW†	TYPES TA, TBS, SA SIS, FEP†, FEPB†, MI RHH†, RHW-2, THHN†, THHW†, THW-2, THWN-2, USE-2, XHH, XHHW† XHHW-2, ZW-2	TYPES TW†, UF†	TYPES RH†, RHW†, THHW†, THW†, THWN†, XHHW†, USE†	TYPES TA, TBS, SA, SIS, THHN†, THHW†, THW-2, THWN-2, RHH†, RHW-2 USE-2 XHH, XHHW XHHW-2, ZW-2	AWG kcmil
	COPPER			ALUMINUM OR COPPER-CLAD ALUMINUM			
18	....	....	14	....	....	....	....
16	....	....	18	....	....	....	....
14	20†	20†	25†	....	....	....	....
12	25†	25†	30†	20†	20†	25†	12
10	30	35†	40†	25	30†	35†	10
8	40	50	55	30	40	45	8
6	55	65	75	40	50	60	6
4	70	85	95	55	65	75	4
3	85	100	110	65	75	85	3
2	95	115	130	75	90	100	2
1	110	130	150	85	100	115	1
1/0	125	150	170	100	120	135	1/0
2/0	145	175	195	115	135	150	2/0
3/0	165	200	225	130	155	175	3/0
4/0	195	230	260	150	180	205	4/0
250		255	170				250

Where conditions are as described in the table heading, then conductor ampacity can be taken to be equal to the table value.

Figure 1

described conditions, the conductor ampacity would be equal to the table value.

If the conditions of use are different from those described by the heading of Table 310-16, then the value of ampacity must be adjusted. The only differences there could be are:

- more than three current-carrying conductors, or
- a higher ambient temperature, or
- both.

And because either of those conditions or the combination of both will cause the conductors to operate at higher temperatures while carrying the same amount of current, the amount of current that the conductors carry must be reduced to prevent the conductor insulation from being damaged.

Procedure 2

Where there are *not* more than three current-carrying conductors in a cable, *but* the ambient temperature is *higher* than 30°C (86°F), then the value shown in Table 310-16 must be adjusted according to the ambient temperature "Correction Factors" given at the bottom of Table 310-16 (Figure 2). The factor selected must be taken from the column that corresponds to the type of insulation used and the actual ambient temperature. For example, if THHN insulation is used and the actual ambient temperature is, say, 33°C, then the value of ampacity shown in Table 310-16 for the size of conductor used must be derated by a factor of 0.96, as shown in Figure 2. That is, in this case, the conductor ampacity equals the Table value for the particular

CORRECTION FACTORS

Ambient Temp. °C	For ambient temperatures other than 30°C (86°F), multiply the allowable ampacities shown above by the appropriate factor shown below.						Ambient Temp. °F
21-25	1.08	1.05	1.04	1.08	1.05	1.04	70-77
26-30	1.00	1.00	1.00	1.00	1.00	1.00	78-86
31-35	.91	.94	.96	.91	.94	.96	87-95
36-40	.82	.88	.91	.82	.88	.91	96-104
41-45	.71	.82	.87	.71	.82	.87	105-113
46-50	.58	.75	.82	.58	.75	.82	114-122
51-55	.41	.67	.76	.41	.67	.76	123-131
56-60	. . . .	.58	.71	. . . .	.58	.71	132-140
61-70	. . . .	.33	.58	. . . .	.33	.58	141-158
71-80	. . . .	. . . .	.41	. . . .	. . . .	.41	159-176

Correction factors for table value where ambient temperature is greater than 30°C (86°F) or lower than 26°C (78°F).

Figure 2

conductor size times 0.96. Generally stated, for ambient temperatures greater than 30°C

Ampacity = Table value × Ambient temperature correction factor

In a like manner, if the ambient temperature is less than 26°C (78°F), then the table value may be increased as indicated by the "Correction Factors" table at the bottom of Table 310-16.

Procedure 3

Where there are *more than* three current-carrying conductors and the ambient temperature is not greater than 30°C (86°F), then the value shown in Table 310-16 must be reduced as required by the Table to Note 8 of the ampacity tables for the actual number of current-carrying conductors within the cable (Figure 3). For example, if there were eight current-carrying conductors, the Table to Note 8 shows us that the value shown in Table 310-16 for the specific insulation type and conductor size(s) within the cable must be reduced by a factor of 0.70, or, for more than three current-carrying conductors

Ampacity = Table 310-16 value × Note 8 correction factor

8. Adjustment Factors.

(a) More than Three Current-Carrying Conductors in a Raceway or Cable. Where the number of current-carrying conductors in a raceway or cable exceeds three, the allowable ampacities shall be reduced as shown in the following table:

Number of Current-Carrying Conductors	Percent of Values in Tables as Adjusted for Ambient Temperature if Necessary
4 through 6	80
7 through 9	70
10 through 20	50
21 through 30	45
31 through 40	40
41 and above	35

Correction factors that must be applied to the ampacity table value where there are more than three current-carrying conductors contained within the raceway, cable, or trench.

Figure 3

Procedure 4

Where there are *more than* three current-carrying conductors *and* the ambient temperature is *higher* than 30°C, then the value of current shown in Table 310-16 for the specific insulation type and conductor size within the cable must be reduced by the factors shown in *both* the ambient temperature "Correction Factor" table (Figure 2) *and* the Table to Note 8 for the actual number of current-carrying conductors (Figure 3). If we had an ambient temperature of 33°C and there were eight current-carrying conductors within the cable, the ampacity would be equal to the Table 310-16 value, for the specific insulation type and conductor size, times the ambient temperature correction factor (0.96) *and* times the Note 8 correction factor (0.70), or Ampacity = Table 310-16 value × 0.96 × 0.70. For both elevated ambient and more than three current-carrying conductors

Ampacity = Table 310-16 value × Ambient temperature correction factor × Note 8 correction factor

In determining such things as box sizes, the neutral conductors are included in the total number of conductors because they occupy space. A completely separate consideration, however, is the relation of neutral conductors to the total number of "current-carrying" conductors, which determines if, and what, derating factor from the Table to Note 8 must be applied.

As covered in Note 10(a) to the ampacity tables (310-16 through 310-19), neutral conductors that are true neutrals—that is, neutrals that carry only the unbalanced current for normally balanced single-phase, 3-wire or 3-phase 4-wire circuits supplying resistive loads—are *not* to be counted with the phase for the purposes of applying Note 8 derating factors. *But,* neutral conductors used with two legs of a 3-phase, 4-wire system, and neutral conductors used with 3-phase, 4-wire circuits that supply electronic discharge lighting, computers, peripherals, faxes, copiers, and similar electronic equipment *must* be counted as current-carrying conductors. Those neutrals are required to be counted because under normally balanced conditions, the neutral will be carrying current that approximately equals, and in some cases exceeds, the phase current. This is a result of additive harmonics on the neutral. Where the load consists primarily of electronic equipment, use of the oversized or individual neutrals on multiwire branch circuits and feeders will assure that the neutral conductors do not become overloaded, which is generally required for all conductors by Sec. 310-10. And such neutrals *would* be counted as current-carrying conductors for the purpose of selecting Note 8 derating factors (Figure 4).

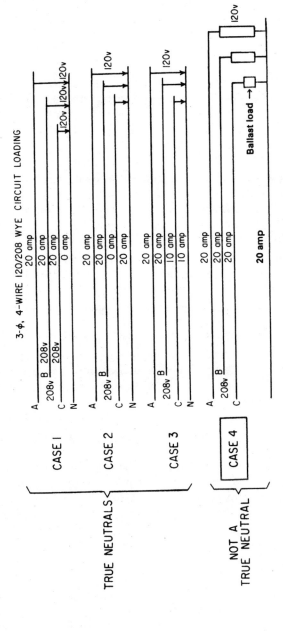

3-φ, 4-WIRE 120/208 WYE CIRCUIT LOADING

CASE 1

CASE 2

CASE 3

TRUE NEUTRALS

CASE 4

NOT A
TRUE NEUTRAL

"True" neutrals are *not* counted when determining the number of
current-carrying conductors (Note 10 to the Ampacity Tables).

Figure 4

Note 11 to the ampacity tables indicates that equipment grounding conductors and bonding conductors are *not* to be counted when applying the rule of Note 8. The reasoning is the same as for true neutrals. That is, although those conductors occupy space, they do not normally carry current and need not be counted.

Overcurrent Protection

After the ampacity of a given conductor has been established in accordance with one of the four procedures described above, a suitably rated overcurrent protective device must be selected to properly protect the conductors. The rating of overcurrent protective devices must be determined in accordance with Sec. 240-3.

The basic rule of that NEC section requires all conductors to be protected by a fuse or CB with a rating equal to the conductor's ampacity. But, because calculated ampacity values do not always directly correspond to the NEC-recognized standard ratings (given in Sec. 240-6), it is permissible to use the next higher standard rated device—as indicated by Sec. 240-6—*provided* the circuit is not rated over 800A *and* the circuit is not part of a multioutlet branch circuit supplying receptacle outlets for cord-and-plug-connected portable loads. For circuits rated over 800A or for multioutlet receptacle circuits, the fuse or circuit breaker must be equal to the conductor's ampacity, *or* the next *lower* standard rating must be used [Secs. 240-3(b) and (c)] (Figure 5).

The remaining parts of Sec. 240-3 [parts (d) through (m)] address those instances where a conductor does *not* have to be protected against overcurrent in accordance with its ampacity. For example, motor branch-circuits and feeders are to be protected as required in Art. 430. Read the various other parts of Sec. 240-3 to assure that the appropriate Code section selected for determining the maximum rating of overcurrent protective device in your particular application is the correct one.

It should be noted that, as indicated by the obelisk (or "dagger") footnote to Table 310-16, No. 14, No. 12, and No. 10 copper conductors may never be protected at more than 15A, 20A, and 30A, respectively. Table 310-16 says THHN insulated, No. 12 copper has an ampacity of 30A. And that value *is* the ampacity of No. 12 THHN copper conductors where used as described in the table heading. However, due to limitations inherent in low-amperage breakers, *any* insulated—TW, THW, THHN, etc.—No. 12 copper conductor must be protected, and therefore loaded, as if it were a 20A wire. *But,* any deratings that may be required for elevated ambient, more than three conductors, or both, may be applied against the 30A value shown in Table 310-16 (Figure 6).

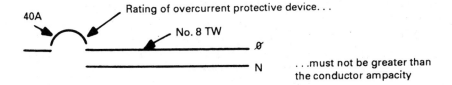

40A

Rating of overcurrent protective device. . .

No. 8 TW

. . .must not be greater than the conductor ampacity

240-3 (b): For circuits rated not over 800A. . .

Rating of overcurrent protective device may be the next higher standard rating above the conductor's ampacity (85A).

90A

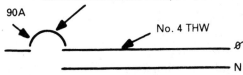

No. 4 THW

Unless the conductors are part of a multioutlet branch circuit supplying cord-and-plug-connected portable loads. In such cases, the next *lower* standard rating from Sec. 240-6 must be used.

240-3 (c): For circuits rated over 800A, the rating of the overcurrent protective device must be equal to or the next lower rating below the conductor ampacity.

Figure 5

Another concern related to the operational characteristics of overcurrent protective devices is whether the load is "continuous" (operates for 3 hours or more without being deenergized) or "noncontinuous" (operates for less than 3 hours) [Art. 100]. In selecting an overcurrent protective device to protect a conductor at a particular ampacity, the

Note to Table 310-16:

†Unless otherwise specifically permitted elsewhere in this Code, the overcurrent protection for conductor types marked with an obelisk (†) shall not exceed 15 amperes for No. 14, 20 amperes for No. 12, and 30 amperes for No. 10 copper; or 15 amperes for No. 12 and 25 amperes for No. 10 aluminum and copper-clad aluminum after any correction factors for ambient temperature and number of conductors have been applied.

Where required to be protected in accordance with their ampacity—as indicated by Sec. 240-3—No. 14, No. 12, and No. 10 must be protected as indicated by this footnote to Table 310-16.

Figure 6

maximum permitted device rating and the conductor ampacity remain the same whether the circuit loading is "continuous" or "noncontinuous." If the load is all "noncontinuous," then the circuit may be loaded up to the rating of the overcurrent protective device. *But,* if all or a portion of the load is "continuous" in nature, then the loading must be limited. That is, the value of noncontinuous load *plus* 1.25% of the continuous load must not exceed the rating of the protective device, unless the device is listed and marked for use at 100% loading for continuous operation. It should be noted that virtually no devices rated 225A, or less, are so listed. Therefore, all overcurrent devices rated 225A, or less, must be evaluated with respect to whether the load is continuous or noncontinuous [Secs. 210-22(c), 220-3(a), and 220-10(c)].

The typical impact of the foregoing rules is illustrated in Figure 7. As can be seen there are two, 3-phase, 4-wire circuits within the raceway supplying fluorescent lighting, which is a harmonic producing, continuous load.

Table 310-16 shows that No. 12 THHN copper conductors have an ampacity of 30A. Normally, the neutrals balanced 3-phase, 4-wire circuits are not counted. But, because Note 10 requires the neutral conductor to be counted where supplying electric discharge lighting, there are eight current-carrying conductors, not six. Therefore, the 70% derating value for 7 to 9 conductors from the Table to Note 8 must

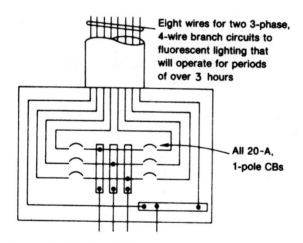

Eight wires for two 3-phase, 4-wire branch circuits to fluorescent lighting that will operate for periods of over 3 hours

All 20-A, 1-pole CBs

With THHN, XHHW or RHH No. 12 copper conductors:
No 12 ampacity is 30 amps from Table 310-16
From Note 8. derating 0.7 X 30 amps = 21 amps
From Sec. 220-3(a). max. load: 0.8 X 20 amps = 16 amps

Figure 7

be applied to the 30A ampacity table value. And the ampacity of the No. 12 conductors is equal to 21A (30A × 0.70). Because the obelisk footnote prohibits protecting a No. 12 copper conductor at more than 20A, a 20A CB is selected.

The maximum amount of continuous load that may be supplied by a 20A protective device is that value that when multiplied by 1.25 will *not* be greater than the rating of the 20A circuit protective device. Therefore, we *divide* the 20A device rating by 1.25 and find that 16A is the value that when *multiplied* (the inverse of division) by 1.25 equals 20A, or

$$20A = 1.25 \times \text{total continuous load}$$

$$20A/1.25 = \text{continuous load}$$

$$16A = \text{continuous load}$$

An example of an application with a mix of continuous and noncontinuous loads is illustrated in Figure 8. As can be seen, where 4A of

Secs. 210-22(c), 220-3(a), and 220-10(c)

require that the rating of the overcurrent protective device be equal to the sum of any noncontinuous load plus 125% of the continuous load.

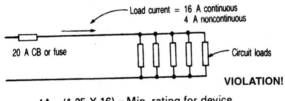

Load current = 16 A continuous
4 A noncontinuous

20 A CB or fuse

Circuit loads

VIOLATION!

4A + (1.25 X 16) = Min. rating for device
4A + 20A = Min. rating for device
24A = Min. rating for device

20A device may *not* supply this combination of load.

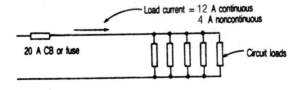

Load current = 12 A continuous
4 A noncontinuous

20 A CB or fuse

Circuit loads

4A + (12A X 1.25) = Min. rating for device
4A + 15A = Min. rating for device
19A = Min. rating for device

20A device here satisfies NEC rules regarding continuous loading of protective devices.

Figure 8

noncontinuous load is to be supplied, the maximum permitted continuous load is 12.8A (16A of remaining circuit capacity divided by 1.25). The first example shown in Figure 8 satisfied the literal wording used in the 1984 NEC and several previous editions, but it was contrary to the intent of the rule. The wording of the related Code rules on continuous loading have since been changed to literally require application as described here.

A Common Controversy:
Bonding of Subpanels

Secs. 250-23 and 250-24. One of the most commonly encountered controversies involved in electrical work is the matter of "bonding" subpanelboards. May a panel on the load side of the service disconnect ever have its neutral busbar "bonded" to the metal enclosure and/or to the equipment grounding busbar? Is such a panel ever required to have its neutral bus bonded?

The sequence of "grounding" and "bonding" considerations that are related to this issue are:

- NEC design rules are concerned with the grounding of a utility-fed AC electric supply circuit that has one of its conductors intentionally grounded—such as a 240/120V single-phase system or 208/120V three-phase, four-wire system or 480/277V three-phase, four-wire system that is required to have its neutral grounded. For such a grounding connection, a grounding-electrode conductor must be connected to the grounded conductor (the neutral conductor) anywhere from the load end of the service drop or lateral to the neutral block or bus within the enclosure for the service disconnect—which includes a meter socket, a CT cabinet, or an auxiliary gutter or other enclosure ahead of the service disconnect, panelboard, or switchboard.

- The grounding-electrode conductor that is connected to the grounded neutral (or grounded phase leg) must be run to a "grounding electrode system" as described in NEC Section 250-81.

- Exception No. 5 of NEC Section 250-23 permits the grounding electrode conductor to be connected to the equipment grounding bus—where such connection is considered necessary to prevent the desensitization of a service ground-fault protection (GFP) hookup that senses fault current through a CT-type sensor on the ground strap between the neutral bus and the ground bus.

The grounding electrode conductor may be connected to either the neutral bus (or terminal lug) or to the equipment grounding block or bus in any system that has a conductor or a busbar bonding the neutral bus or terminal to the equipment grounding block or bus. But, where the neutral is bonded to the enclosure simply by a bonding screw, the grounding electrode conductor must *not* be connected to the equipment grounding block or bus. It must be connected only to the neutral-conductor terminal block or bus. This rule is based on the idea that a lightning strike on the neutral conductor outside would cause the lightning current to flow through the bonding screw in going to ground through a grounding-electrode conductor that is connected to the equipment ground terminal. The conductivity of a screw-type bond is not adequate to effectively pass the very high lightning currents to earth.

One of the most important and widely discussed considerations in the design of electrical distribution systems is the matter of making a grounding connection (or equipment bond) to the system grounded neutral or grounded phase wire. The NEC says "A grounding connection shall not be made to any grounded circuit conductor on the load side of the service disconnecting means." [Sec. 250-23(a)]. Once a neutral or other circuit conductor is connected to a grounding electrode at the service equipment, the general rule is that the neutral or other grounded leg must be insulated from all equipment enclosures and other grounded parts on the load side of the service. That is, bonding of subpanels (or any other connection between the neutral or other grounded conductor and equipment enclosures) is prohibited.

There are some exceptions to that rule, but they are few and are very specific:

- In a system, even though it is on the load side of the service, when voltage is stepped down by a transformer, a grounding connection must be made to the secondary neutral of a 208/120V or 480/277V system;

- When a circuit is run from one building to another, it may be necessary (or simply permissible) to connect the system "grounded" conductor to a grounding electrode and to bond the neutral to the equipment ground bus; and

- The frames of ranges, wall ovens, countertop cooking units, and clothes dryers may be "grounded" by connection to the grounded neutral of their supply circuit (Section 250-60).

Within a building, it is a clear violation to bond the neutral block in a panelboard to the panel enclosure in other than a service panel. In a panelboard used as service equipment, the neutral block (terminal block) is bonded to the panel cabinet by the bonding screw provided. And such bonding is required to tie the grounded conductor to the interconnected system of metal enclosures for the system (i.e., service equipment enclosures, conduits, busways, boxes, panel cabinets). It is this connection that provides for the flow of fault current and operation of the overcurrent device (fuse or circuit breaker) when a ground-fault occurs. But, there must not be any connection between the grounded system conductor and the grounded metal enclosure system at any point on the load side of the service equipment, because such connection would constitute connection of the grounded system conductor to a grounding electrode (through the enclosure and raceway system to the water pipe or driven ground rod). Such connections, like the bonding of subpanels, can be dangerous, as shown in Figure 1.

The NEC rule prohibiting connection of the grounded system wire to a grounding electrode on the load side of the service disconnect must not be confused with the rule of Section 250-60, which permits the grounded system conductor to be used for grounding the frames of electric ranges, wall ovens, counter-mounted cooking units, and electric

1. THIS CONDITION WILL EXIST....AND...

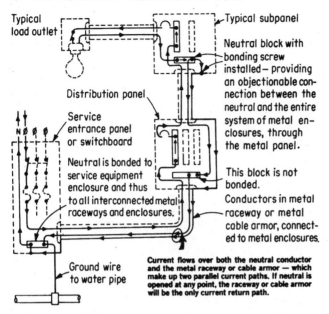

Typical load outlet

Typical subpanel

Distribution panel

Neutral block with bonding screw installed — providing an objectionable connection between the neutral and the entire system of metal enclosures, through the metal panel.

Service entrance panel or switchboard

Neutral is bonded to service equipment enclosure and thus to all interconnected metal raceways and enclosures.

This block is not bonded.

Conductors in metal raceway or metal cable armor, connected to metal enclosures.

Ground wire to water pipe

Current flows over both the neutral conductor and the metal raceway or cable armor — which make up two parallel current paths. If neutral is opened at any point, the raceway or cable armor will be the only current return path.

2. THIS HAZARD COULD DEVELOP

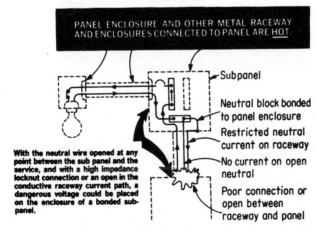

PANEL ENCLOSURE AND OTHER METAL RACEWAY AND ENCLOSURES CONNECTED TO PANEL ARE <u>HOT</u>.

Subpanel

Neutral block bonded to panel enclosure

Restricted neutral current on raceway

No current on open neutral

With the neutral wire opened at any point between the sub panel and the service, and with a high impedance locknut connection or an open in the conductive raceway current path, a dangerous voltage could be placed on the enclosure of a bonded subpanel.

Poor connection or open between raceway and panel

Figure 1 Bonding the neutral block to the panel enclosure in other than a service panel is a clear violation of the NEC. The reason for this restriction is to prevent the hazardous condition illustrated here.

clothes dryers. The connection referred to in Section 250-60 is that of an ungrounded metal enclosure to the grounded conductor for the purpose of grounding the enclosure.

Grounding at Separate Buildings

As described above, bonding of a panel neutral block (or the neutral bus or terminal in a switchboard, switch, or circuit breaker) to the enclo-

For grounded systems . . .

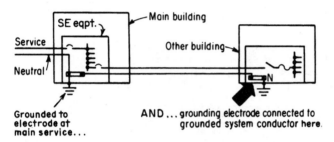

Grounded to electrode at main service...

AND ... grounding electrode connected to grounded system conductor here.

. . . and ungrounded systems.

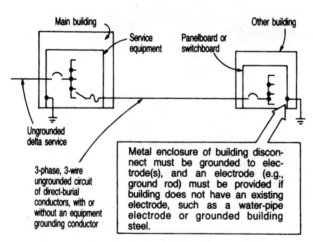

Figure 2 Section 250-24(a) requires that when two or more buildings are supplied from a single installation of service equipment in one of the buildings, a grounding electrode at each building must be connected to the ac system grounded conductor on the supply side of a building feeder disconnecting means of a grounded system, or connected to the metal enclosures of the building disconnecting means of an ungrounded system.

sure is required by the NEC in service equipment. But the NEC permits bonding of the neutral conductor on the load side of the service equipment in those cases where a panelboard (or switchboard, switch, etc.) is used to supply circuits in one building and the panel is fed from a service in another building.

Where two or more buildings are supplied from a single installation of service equipment in one of the buildings, a grounding electrode at each building must be connected to the AC system grounded conductor on the supply side of the building feeder disconnecting means of a grounded system, or connected to the metal enclosures of the building disconnecting means of an ungrounded system (Figure 2).

That is the basic rule. But an exception to the rule notes that a grounding electrode at a separate building supplied by a feeder or branch circuit is not required where either of the following conditions is met:

1. Only one branch circuit is supplied, and there is no noncurrent-carrying equipment in the building that requires grounding (such as a small toolshed with a single lighting outlet or switch with no metal boxes, or faceplates, or there are no lighting fixtures within eight feet vertically or five feet horizontally from any grounded object).

2. An "equipment grounding conductor" is run with the circuit conductors for the grounding of noncurrent-carrying equipment, water piping, or building metal frames in the separate building, and no livestock are housed in the building (Figure 3).

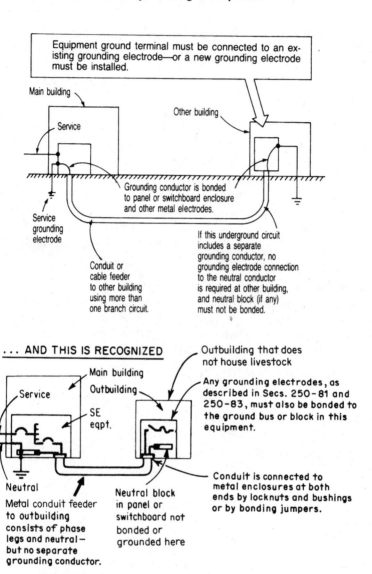

Figure 3 Alternate method for grounding of electrical system supplying more than one building. Notice that connection to a new or existing grounding electrode system is still required.

NEC Rules for...
Cables Run Through, or Parallel to, Metal and Wood Framing Members

Sec. 300-4. **In certain applications the** NEC **requires that cables be provided with additional protection to guard against physical damage to the cable assembly. That is, due to the increased hazards associated with installing cable in some locations, supplemental protection for the cable must be field installed.**

Sec. 300-4(a) gives the rules on the protection required where cables are run through holes or notches in wood framing members (Figure 1). Where the edge of the hole in a wood member is less than 1-1/4 inch from the nearest edge of the member, a 1/16 in-thick steel plate must be used to protect the cable sheath against penetration by driven nails or screws. The same protection is required where cable is laid in a notch in the wood member.

Clearances must be provided from both edges of the hole in a wood member to both edges of the wood member if it is likely that a nail or screw may be driven from either side. Where the NEC used to require a 1-1/2 inch clearance from the edge of the hole to the nearest edge of the stud, the present NEC only requires a clearance of 1-1/4 inch. This permits realistic compliance when drilling holes through studs that are 3-1/2-inch deep. It also was taken into consideration that the nails commonly used to attach wall surfaces to studs were of such a length that the 1-1/4-inch clearance to the edge of the hole afforded entirely adequate protection against penetration of the cable by the nail.

The rule of Sec. 300-4(d) was new to the Code in 1990 and has proved to be extremely controversial where Type AC or MC cables are used. This section requires cables run along (parallel to) metal or wood framing members to have at least a 1-1/4-inch clearance from the nearest edge; otherwise the cable must be protected by a 1/16-inch-thick steel plate or sleeve (bottom of Figure 1). And this applies to either concealed or exposed work. This rule was added because of the many persistent reports of nail and screw penetration of nonmetallic cables, as well as some nonmetallic and metallic raceways. However, no reports were made of any such problem with Type AC or MC. Regardless, the literal wording of the rule simply calls for additional protection where "cables" are run parallel to wood or metal framing members, and no exception is given for Type AC or MC except for concealed work where the cable is fished [Sec. 300-4(d), Ex. No. 2].

If BX, NM cable, or raceway wiring (rigid conduit, EMT, etc.) is used through holes bored in joists, rafters or similar wood members. . .

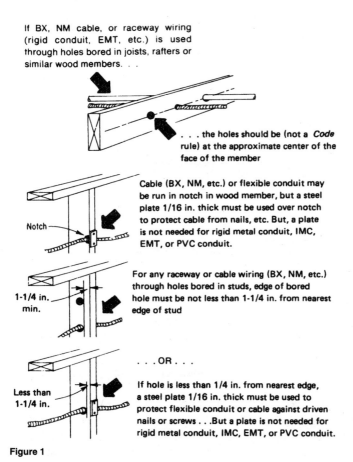

. . . the holes should be (not a *Code* rule) at the approximate center of the face of the member

Cable (BX, NM, etc.) or flexible conduit may be run in notch in wood member, but a steel plate 1/16 in. thick must be used over notch to protect cable from nails, etc. But, a plate is not needed for rigid metal conduit, IMC, EMT, or PVC conduit.

Notch

For any raceway or cable wiring (BX, NM, etc.) through holes bored in studs, edge of bored hole must be not less than 1-1/4 in. from nearest edge of stud

1-1/4 in. min.

. . . OR . . .

Less than 1-1/4 in.

If hole is less than 1/4 in. from nearest edge, a steel plate 1/16 in. thick must be used to protect flexible conduit or cable against driven nails or screws . . .But a plate is not needed for rigid metal conduit, IMC, EMT, or PVC conduit.

Figure 1

It should be noted that prior to the release of the 1993 NEC a Tentative Interim Amendment (TIA) was issued to exempt Type AC and Type MC cable from the need for supplemental physical protection where run parallel to framing members. Additionally, a proposal was submitted and accepted by the Code Making Panel to exclude Type AC and Type MC from the rule of Sec. 300-4(d). And that action was supported by public comments submitted during the comment period for the 1993 NEC. However, through some mysterious sequence of events, the exemption accepted was not reflected in the final version of the 1993 NEC.

This rule is still widely viewed as excessive where Type AC or MC cable is run along the sides of metal or wood members. In fact, the state

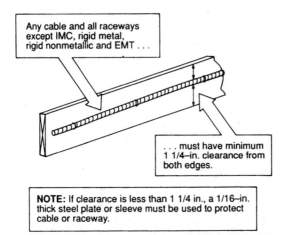

Any cable and all raceways except IMC, rigid metal, rigid nonmetallic and EMT . . .

. . . must have minimum 1 1/4–in. clearance from both edges.

NOTE: If clearance is less than 1 1/4 in., a 1/16–in. thick steel plate or sleeve must be used to protect cable or raceway.

Wiring on structural members must have clearance for protection against nails.

Figure 2

of Massachusetts waived the rule of Sec. 300-4(d) for Type AC and MC in addition to those other raceways mentioned in Ex. No. 1. There are other jurisdictions where the requirement for protection of AC and MC in such application has also been suspended. Therefore, it is best to check with the local electrical inspector to determine if the clearances described in Sec. 300-4(d) apply to Type AC or MC cable (Figure 2).

In the event the rule of Sec. 300-4(d) is in effect, there are commercially available special clamps that will serve to satisfy the clearance requirement and eliminate the need for sleeving. These devices are mounted on the framing member and can support a number of Type AC or MC cables at the prescribed distance from the edge.

1993 NEC Rules for Motor Short-Circuit and Ground-Fault Protection

Secs. 430-52(a), 430-62(a), and Exceptions. Subtle changes in wording dramatically alter basic procedures for determining the maximum rating of overcurrent protective devices on motor branch-circuits and feeders.

Of the many changes contained in the 1993 edition of the National Electrical Code (NEC), the revised wording of the rules for "short-circuit and ground-fault protection" of motor branch circuits and feeders represent a rather significant departure from previously accepted practice. These newly revised rules greatly complicate and confuse the basic everyday task of selecting the maximum allowable rating or setting for a motor-circuit overcurrent protective device as covered in Secs. 430-52, 430-62, and their Exceptions. Additionally, the new wording leaves questions as to what is, and what is not, acceptable in certain applications.

The most disturbing aspect of most of these changes is that no real benefit has been achieved. That is, although the basic thrust of the revisions is to provide for closer protection of motor circuits, no one indicated that rules for motor circuit overcurrent protection as given in the 1990 NEC presented any compromise in safety or increased hazard. Not once throughout all discussions in the Technical Committee Report and the Technical Committee Documentation was any documentation or comment offered to indicate that an actual problem existed in application, which would truly justify the accepted revisions.

Although these changes may initially present an unwelcomed additional burden for designers and installers, everyone must resign themselves to the fact that compliance with the modified requirements is *not* optional; it is mandatory. The following discussion and analysis explains what is now required, identifies potential problem applications, and offers considered interpretations that should assure safety, while accommodating practical application.

Individual Motor Branch Circuit Protection

Part D of Art. 430 presents rules for "short-circuit and ground-fault protection" of motor circuits. Although the phrase "ground-fault protection" is used in several sections of Part D, it should be noted that it refers to the protection against ground-fault that is provided by the fuses, CBs, MSCP, or Instantaneous-Trip CBs that also provide short-circuit protection. These rules are *not* intended to require the type of ground-fault protective hookup required for service disconnects, main building disconnects, and feeder disconnects (Secs. 230-95, 240-13, and 215-10, respectively; see Figure 1).

The Code requires that branch-circuit protection for motor circuits must protect the circuit conductors, the control apparatus, and the motor itself against overcurrent due to short circuits and ground faults

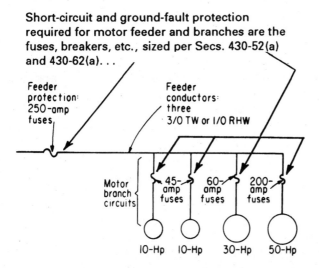

Short-circuit and ground-fault protection required for motor feeder and branches are the fuses, breakers, etc., sized per Secs. 430-52(a) and 430-62(a)...

Feeder protection: 250-amp fuses

Feeder conductors: three 3/0 TW or 1/0 RHW

Motor branch circuits

45-amp fuses   60-amp fuses   200-amp fuses

10-Hp   10-Hp   30-Hp   50-Hp

...*not* the type of hookup required for high-capacity service, feeder, and main building disconnects (Secs. 230-95, 215-10, and 240-13, respectively), shown below.

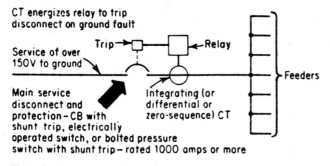

CT energizes relay to trip disconnect on ground fault

Trip →   ← Relay

Service of over 150V to ground

Feeders

Main service disconnect and protection-CB with shunt trip, electrically operated switch, or bolted pressure switch with shunt trip—rated 1000 amps or more

Integrating (or differential or zero-sequence) CT

Figure 1

(Secs. 430-51 through 430-58). The first, and obviously necessary, rule is that the branch-circuit protective device must be capable of carrying the starting current of the motor without opening the circuit. Then the Code proceeds to place maximum values on the settings or ratings of such overcurrent protective devices. The maximum rating or setting of a short-circuit and ground-fault protective device for other than torque motors is established in accordance with Sec. 430-52(a).

The basic requirement for individual motor branch circuit short-circuit and ground-fault protection remains unchanged in the 1993 NEC. As stated in Sec. 430-52(a), the maximum rating or setting of the short-circuit and ground-fault protective device—fuse, CB, etc.—must not exceed the percentage of motor full-load current indicated by Table 430-152 for the specific type of device selected (Figure 2).

As can be seen, for all AC single- and poly-phase squirrel-cage motors with no code letter and for synchronous motors with no code letter [other than low-speed (450rpm or less), low-torque type synchronous motors], if a nontime-delay fuse is used to provide short-circuit and ground-fault protection, then the rating of that fuse must not exceed a value that is equal to 300% of (3 times) the motor full-load current. Where a time-delay fuse is used, the rating must not exceed 175% of (1.75 times) the motor full-load current. Instantaneous-trip CBs are limited to a maximum setting of 700% (7 times) motor full-load current, and conventional inverse-time CBs must be rated or set at no more than 250% (2.5 times) the motor full-load current.

The first change to be discussed relates to Exception No. 1, which covers those cases where the value calculated in accordance with the basic rule does *not* correspond to a standard rating or possible setting. In the 1990 and many previous editions of the Code, Sec. 430-52(a), Exception No. 1 was widely interpreted to permit selection of the next *higher* rating or possible setting if the calculated value did not correspond to one of the Code-recognized standard ratings given in Sec. 240-6. The wording of Exception No. 1 in the 1993 NEC now clearly requires selection of the next *lower* standard rating or setting, unless the next lower rating or setting is inadequate to carry the starting current.

This whole matter can be traced back to a proposal submitted for the 1987 NEC that attempted to rectify the apparent lack of correlation between Example No. 8 in Chapter 9 and the generally accepted interpretation of Ex. No. 1 to Sec. 430-52(b). The proposal (No. 11-113, '86 TCR) requested that Example No. 8 be revised to reflect the commonly accepted practice of rounding up where the calculated maximum value did not correspond to a standard device rating or possible setting. That proposal was rejected because the Code Making Panel (CMP) stated its intent was to require rounding down in such cases.

Another proposal was submitted for the 1990 NEC to clarify the word-

430-152.   Maximum Rating or Setting of Motor Branch-Circuit Short-Circuit and Ground-Fault Protective Devices

	Percent of Full-Load Current			
Type of Motor	Nontime Delay Fuse	Dual Element (Time-Delay) Fuse	Instan-taneous Trip Breaker	Inverse Time Breaker*
Single-phase, all types				
No code letter	300	175	700	250
All ac single-phase and polyphase squirrel-cage and synchronous motors† with full-voltage, resistor or reactor starting:				
No code letter	300	175	700	250
Code letters F to V	300	175	700	250
Code letters B to E	250	175	700	200
Code letter A	150	150	700	150
All ac squirrel-cage and synchronous motors† with autotransformer starting:				
Not more than 30 amps				
No code letter	250	175	700	200
More than 30 amps				
No code letter	200	175	700	200
Code letters F to V	250	175	700	200
Code letters B to E	200	175	700	200
Code letter A	150	150	700	150
High reactance squirrel-cage				
Not more than 30 amps				
No code letter	250	175	700	250
More than 30 amps				
No code letter	200	175	700	200
Wound-rotor —				
No code letter	150	150	700	150
Direct-current (constant voltage)				
No more than 50 hp				
No code letter	150	150	250	150
More than 50 hp				
No code letter	150	150	175	150

For explanation of code letter marking, see Table 430-7(b).

For certain exceptions to the values specified, see Sections 430-52 through 430-54.

*The values given in the last column also cover the ratings of nonadjustable inverse time types of circuit breakers that may be modified as in Section 430-52.

†Synchronous motors of the low-torque, low-speed type (usually 450 rpm or lower), such as used to drive reciprocating compressors, pumps, etc., that start unloaded, do not require a fuse rating or circuit-breaker setting in excess of 200 percent of full-load current.

Figure 2

ing of NEC rules for short-circuit and ground-fault protection as covered in Art. 430 by more clearly indicating the CMP's intent as stated in the '86 TCR. The substantiation for that proposal (11-37 in the '89 TCR) explicitly pointed out that Sec. 430-52(a), Ex. No. 1 was consistently interpreted to permit rounding up where the calculated value did not correspond to one of the standard ratings given in Sec. 240-6. Curiously enough, and with virtually no explanation, the CMP rejected the proposal and again stated that rounding down was the intended requirement. Such action is completely mystifying because the net result was CMP rejection of a concept with which it agreed! That reality was brought to light when the concern for clarification of the CMP's intent was again the subject of a proposal for the 1993 NEC. This time, the CMP decided to accept a change in wording that clearly calls for rounding down, unless the lower standard rating or setting will not be capable of carrying the starting current.

This whole thing seems to have gotten out of hand. What began as a request to coordinate Example No. 8 with commonly accepted application evolved into a radical modification of standard practice. And such modification ignores the thousands upon thousands of installations that have been made using the previously accepted interpretation without any report of incident or accident. Although the basic idea is to provide overcurrent protection rated no higher than absolutely necessary, the present wording of Sec. 430-52(a), Ex. No. 1 complicates the task of selecting a properly rated device beyond that which has been demonstrated as necessary through actual field application. However, to satisfy present NEC requirements, where the calculated maximum does not directly match a standard rating, the next *lower* rating or possible setting must be used, unless the lower rated device is "not adequate to carry" the starting current.

For example, consider the motor branch circuit shown in Figure 3. Where nontime-delay fuses are used to provide the Code-required protection, Table 430-152 shows that the fuses must be rated at no more than 300% (3 times) the motor full-load current. As required by Sec. 430-6, the value of motor full-load current used for the calculation must be taken from Tables 430-147 through 430-150—not the motor nameplate—when establishing the maximum rating of motor short-circuit and ground-fault protection. For the 3-phase, 230V, 50hp motor in question, refer to Table 430-150. That table shows the full-load current must be taken as 130A. Executing the required calculation it can be seen that the one-time (nontime-delay) fuses must be rated at not more than 390A (130A × 3). However, it is not possible to get a nontime-delay fuse rated at 390A. And the new wording of Ex. No. 1 would *not* permit the use of a 400A fuse—the next higher rating—and *requires* that a 350A fuse—the next lower standard rating—be used.

As indicated, Ex. No. 1 would recognize use of the next higher rated 400A fuse provided the next lower rated 350A fuse is not adequate to carry the starting current of the motor. But, how does one go about establishing that fact? The answer to that question is not easy, but the

Sec. 430-52(a) and Ex. No. 1

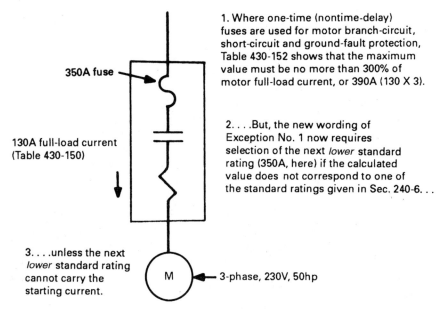

350A fuse

130A full-load current (Table 430-150)

3. . . . .unless the next *lower* standard rating cannot carry the starting current.

1. Where one-time (nontime-delay) fuses are used for motor branch-circuit, short-circuit and ground-fault protection, Table 430-152 shows that the maximum value must be no more than 300% of motor full-load current, or 390A (130 X 3).

2. . . . .But, the new wording of Exception No. 1 now requires selection of the next *lower* standard rating (350A, here) if the calculated value does not correspond to one of the standard ratings given in Sec. 240-6. . . .

M   3-phase, 230V, 50hp

Figure 3

revised wording of the Exception for instantaneous-trip CBs [at the end of Sec. 430-52(a)] seems to provide some guidance.

The Exception following the rules for instantaneous-trip CBs in Sec. 430-52(a) was modified in the 1993 NEC to require that instantaneous-trip CBs be treated the same way as fuses and inverse-time CBs. That is, settings higher than that indicated by Table 430-152 may be used *only* if the indicated maximum setting (700%) causes nuisance-tripping on startup. The last two sentences of that Exception read as follows:

> Trip settings above 700 percent shall be permitted where the need has been demonstrated by engineering evaluation. In such cases, it shall not be necessary to first apply an instantaneous trip circuit breaker at 700%.

The first sentence recognizes "engineering evaluation" as a means of establishing the need for the next higher setting. And the second sentence alludes to actual installation of the device—a sort of on-line test—to determine its adequacy. Use of either of those methods should, logically, be viewed as acceptable methods to establish the need for a higher rating of fuse or CB.

Some may say, "Wait a minute! That wording is *not* included in Exception No. 1! It only applies to instantaneous-trip CBs!"

It is true that the permission given to use "engineering evaluation" literally applies only to instantaneous-trip CBs and is *not* specifically recognized by Exception No. 1. And, because the wording in Sec. 430-52(a) only literally exempts instantaneous-trip CBs from the need to "first apply" the device, that second sentence could be interpreted to imply that all other devices must be installed and "tested." However, such interpretation, in addition to being extremely rigid, fails to recognize certain realities.

Obviously, in new construction projects, as well as many renovation or modernization projects, even if the motor is in place, there generally is no power available to "performance test" the overcurrent device. And, even if power is available, such a "test" may not be conclusive since the amount of starting current will vary. That is, if the circuit is *not* closed on the point in the voltage waveform that creates maximum asymmetry and, as a result, maximum current flow, then it may appear as if the device is adequate. Indeed, the circuit may be closed several times successfully without nuisance tripping, and then, the next time the motor is started, the device may trip or blow and open the circuit.

While everyone would agree that opening of the circuit by the overcurrent protective device upon startup would be a conclusive indication of the need for the next higher rating and may be useful in some instances, such an approach should hardly be considered as the preferred method. And, if "engineering evaluation" is recognized as valid for installations protected by instantaneous-trip breakers, similar evaluation of installations utilizing other devices should also be considered as confirmation of the need for the next higher rating or setting. But just what is "engineering evaluation"?

The term "engineering evaluation" is not defined anywhere in the NEC, but discussions in the TCR and TCD regarding this term allude to a comparison of the time-current trip characteristics of the protective device and the motor in-rush current—including maximum asymmetry—and presumably, other variables, such as load torque. Although this method is preferable to installing the device and energizing the circuit, such an analysis is no easy task.

For the immediate future, designers and installers are going to require additional assistance from manufacturers' sales representatives and application engineering personnel in the form of pertinent technical data, as well as guidance for proper application of their products. As designers and installers become more familiar with the relevant operating characteristics of different manufacturers' equipment, the process of selecting the correct rating to satisfy NEC rules and provide a functional installation should become easier.

Another change in wording that presents a potential problem appears in the basic rule governing settings for instantaneous-trip breakers. In an attempt to reinforce the absolute 1300% maximum setting for instantaneous-trip CBs, the phrase *"and if it will operate at not more than 1300 percent of full-load motor current"* was added at the end of the first sentence. That new wording seemingly *prohibits* the use of any instantaneous-trip CB combination motor controller if it can be set at more than the 13-times maximum—that is, if it will operate at *more than* 1300 percent of full-load current. However, as indicated, the intent behind this change was to emphasize the maximum permitted setting—it was *not* intended to describe the type of device permitted.

Although the rule should *not* be interpreted to prohibit the use of devices that have possible settings above the 1300 percent maximum, the literal requirement must also be addressed to prevent legal exposure. Analysis of rules governing general application of similar equipment provides interesting insight as to how the literal wording may be satisfied.

General application of adjustable-trip CBs is covered in Sec. 240-6(b). The basic rule requires that the rating of adjustable-trip breakers must be taken to be the maximum possible setting. As a result of that rule and the rules for conductor protection (Sec. 240-3), the circuit conductors connected to the load terminals of such a breaker must be of sufficient ampacity to be properly protected at the maximum current value to which the adjustable-trip might be set. That means the CB maximum rating must not exceed the ampacity of the circuit conductors, except where the ampacity of the conductors does not correspond to a standard rating; then, the next higher standard rating may be used for any circuit—other than a multioutlet receptacle circuit—rated up to 800A.

Prior to the 1987 edition, the NEC did not require a circuit breaker with adjustable or changeable trip rating to have load-circuit conductors of an ampacity at least equal to the highest trip rating at which the breaker might be set. Conductors of an ampacity less than the highest-possible trip setting could be used provided the actual trip setting protected the conductors in accordance with its ampacity, as required in Sec. 240-3.

In the '87 NEC, a paragraph was added to Sec. 240-6 to clarify how an adjustable-trip breaker is to be rated. And the accepted wording literally required that adjustable-trip CBs be rated—and therefore cabled—at the maximum possible setting, and *not* the actual setting used. However, the CMP did not really object to the use of adjustable-trip CBs cabled for the actual setting, which had been common practice prior to the '87 Code. Apparently, the CMP was more concerned with the potential for unauthorized or inadvertent tampering of the setting. In response to a number of proposals, the CMP's position was clarified in an Exception that was added to Sec. 240-6 in the 1990 NEC. The Substantiation for Proposal No. 4-186 in the '86 TCR, which added that Exception, in part, stated:

> It is the Panel's intent to permit the use of adjustable trip circuit breakers provided access for adjustment is limited.

And the wording of the Exception indicates what means are recognized as providing the desired limited access. That is, adjustable-trip CBs are permitted to be rated at the actual trip setting provided the breakers "have removable and sealable covers over the adjusting means, or are located behind bolted equipment enclosure doors, or are located behind locked doors accessible only to qualified personnel." (See Figure 4.)

Inasmuch as the Code recognizes the actual setting as the device rating where such precautions are taken in general application of adjustable-trip breakers, the actual setting should also be recognized as the device rating for any listed instantaneous-trip CB combination starter installed with similarly limited access.

Because these type controllers have been in use for a number of years without any reported compromise in safety or operating integrity, and because the Code recognizes the actual trip value as the device rating for general application of adjustable-trip CBs where access is limited, if an instantaneous-trip CB combination starter is set to operate at not more than 13 times full-load current and the adjusting means is sealed

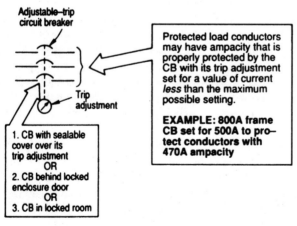

Adjustable–trip circuit breaker

Trip adjustment

1. CB with sealable cover over its trip adjustment
   OR
2. CB behind locked enclosure door
   OR
3. CB in locked room

Protected load conductors may have ampacity that is properly protected by the CB with its trip adjustment set for a value of current *less* than the maximum possible setting.

EXAMPLE: 800A frame CB set for 500A to pro– tect conductors with 470A ampacity

Figure 4

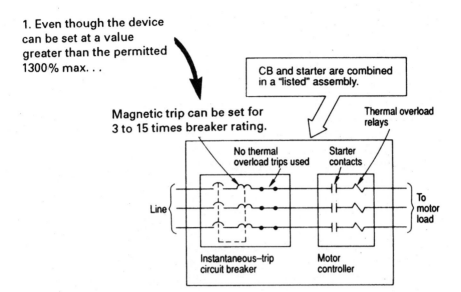

1. Even though the device can be set at a value greater than the permitted 1300% max. . .

CB and starter are combined in a "listed" assembly.

Magnetic trip can be set for 3 to 15 times breaker rating.

Thermal overload relays

No thermal overload trips used

Starter contacts

Line

To motor load

Instantaneous–trip circuit breaker

Motor controller

2. . . . . if such a device were set at "not more than 1300 percent" of full-load and installed with access limited as described in Sec. 240-6(b), use of such a device should be viewed as satisfying the requirement of Sec. 430-52(a).

Figure 5

or installed behind locked or bolted enclosure doors, or in locked rooms, the literal requirement for operation "at not more than 1300 percent" can be reasonably assured and such application should be considered as acceptable.

Important: It must be noted that a setting in excess of 13 times full-load current would be a clear, direct violation of both the "letter" and the "spirit" of this rule (Figure 5).

Motor Feeder Circuit Protection

The rules governing maximum ratings for motor feeder short-circuit and ground-fault protection are given in Part E of Art. 430 (Secs. 430-61 through 430-63). For the 1993 NEC, there are two significant changes to Sec. 430-62(a), which covers the rating of feeder protective devices where the load consists of motors only.

The first significant change is the additional wording provided in the parentheses of the basic rule in Sec. 430-62(a). The new phraseology was added to clarify the CMP's intent regarding the maximum rating permitted for feeder protection where the highest-rated branch-circuit protective device is set or rated at some value *less* than that recognized by Table 430-152. Such applications have been the source of controversy for a number of code cycles because the literal wording was ambiguous.

As most are aware, the rule of Sec. 430-62(a) had required—and still does require—that the maximum rating or setting of the feeder protection be no greater than the sum of the highest-rated branch-circuit protective device plus the full-load current value of all other motors supplied by the feeder. However, in the 1990 and previous editions of the NEC, while the first paragraph of Sec. 430-62 indicated that the ampere value of the highest-rated branch-circuit device was to be based on the maximum given in Table 430-152, the second paragraph referred to the actual value "used." As a result, some believed that it was permissible to use the maximum ampere value permitted by the table and others said the actual rating or setting of the installed device must be used to establish the maximum rating or setting for the feeder protective device.

The problem presented by that lack of clarity is illuminated by the question presented in the Substantiation submitted for Comment No. 11-30 in the '92 TCD. The relevant portion of that substantiation reads as follows:

> Three motors with FLA (full-load amperage) of 40amps are installed each on a separate (branch) circuit with an 80amp breaker. The installer decided not to protect at the maximum 100A permitted (250% FLA for CBs) by 430-52 (and Table 430-152).
>
> The question is: What is the maximum feeder breaker rating permitted? Is it: 80A + 40A + 40A = 160 or 150A breaker or 100A + 40A + 40A = 180 or a 175A breaker? (See Figure 6.)

As can be seen, if the actual installed rating is used, a 150A breaker would be the maximum permitted. And, if the table-permitted maximum value is used, then a 175A breaker would be acceptable.

Although the second paragraph of Sec. 430-62(a) still refers to the "rating or setting...used," inclusion in the first paragraph of the phrase "the maximum permitted value for the specific type of protective device shown" is intended to convey that use of the table value is permitted. That is, it is the CMP's intent to allow use of the 100A rating (250% times 40A) permitted by Table 430-152 to determine the maximum rating or setting of the feeder protection, even though an 80A-rated CB was actually installed. And, although use of the installed 80A rating to establish the maximum rating of feeder protective device would not be prohibited because it results in the selection of a lower value, it is *not* required.

That concept would also be applicable to those cases where the value of branch-circuit protection is "rounded down."

Although the rule of Sec. 430-52(a) and Table 430-152 would permit the feeder breaker to be rated at 100A, the actual rating of the installed device is 80A.

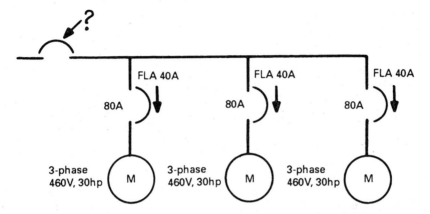

Question: Is the maximum permitted rating for the feeder:
(80A + 40A + 40A), or 160A, which would call for a 150A CB?

OR

is it (100A + 40A + 40A), or 180A, which would call for a 175A CB?

Figure 6

Consider a feeder supplying two 50hp, 3-phase, 230V, motors. Assume that the "next lower standard rating" is adequate to carry the starting current and the branch-circuit fuses are rated as shown in Figure 7. The maximum permitted rating of protective device for a feeder supplying two such motors would be:

Max Feeder Protection Rating = 390A (max permitted by Table 430-152) + 130A

Max Feeder Protection Rating = 520A

It is worth noting that selection of the next *lower* standard-rated device has long been required for motor feeder protection. Therefore, because it is not possible to get 520A fuses, 500A-rated devices would be selected.

The second major change to the procedure for establishing the maximum rating for motor feeder protection is the new exception following Sec. 430-62(a). The new requirement addresses an inherent inadequacy of the basic rule that results in especially excessive ratings for feed-

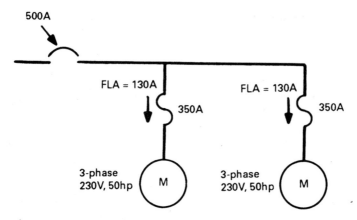

New wording of the basic rule in Sec. 430-62(a) permits use of the maximum value indicated by Table 430-152. In this case, the feeder protective device rating would be:

Max rating = 390A (130 X 3 from Table 430-152) + 130A
           = 520A, or 500A, which is the next lower standard rating.

Figure 7

er protection where instantaneous-trip CBs or MSCPs are used on any branch circuit(s) supplied by the feeder.

As we have seen, the rating of a motor branch-circuit protective device is based on the *type* of device to be used. That is, if one-time fuses are used, they must generally be rated at no more than 300% of motor full-load current; if an inverse-time CB is used, then the multiplier is 250%; for time-delay fuses, the maximum is 175%, etc. However, the basic rule for determining the maximum rating or setting for motor feeder short-circuit and ground-fault protection is based on the rating or setting of the largest branch-circuit protective device—without regard for the type of device on either the feeder or the branch. And that can result in undesirably high or low ratings where certain combinations of devices are used.

For example, consider an application using time-delay fuses to protect a feeder supplying two motor branch-circuits protected by one-time fuses. For the ease of explanation, assume the full-load current for one motor is 100A and the other is 5A. In accordance with Sec. 430-52(a) and Table 430-152, the one-time fuses on the branch may be rated at 300% of motor full-load current. Therefore, in this example, the branch fuses protecting the 100A motor should be rated at not more than 300A (100A × 3). And, the one-time fuses for the 5A motor should be rated at 15A (5A × 3).

In accordance with Sec. 430-62(a), the time-delay fuses would also be permitted to be rated at 300A because the sum of the largest device rating plus the full-load current value of all other motors equals 305A (300A + 5A), which necessitates the use of 300A fuses—the next lower standard rating. Therefore, the one-time fuses on the branch *and* the time-delay fuses on the feeder would be rated at 300A.

While the established maximum ratings satisfy the literal wording, the feeder protection is probably rated too high. As we know, time-delay fuses will carry a higher value of current for a longer period of time than one-time fuses of the same ampere-rating. Therefore, use of a lower ampere-rating for the time-delay fuses on the feeder will assure that the feeder conductors are more closely protected and is probably warranted. The need for lower ratings where time-delay fuses are used is effectively acknowledged by the Code, which logically establishes a lower maximum rating for time-delay fuses (175% of motor full-load current) than for one-time fuses (300%) where used to provide motor branch-circuit protection.

The inherent deficiency of this Code procedure can be more fully appreciated when one considers reversing the position of the different fuse types in the above example. That is, if time-delay fuses were used on the *branch* circuit and one-time fuses were installed on the *feeder,* the maximum rating for the branch and feeder devices would be as follows:

Per Sec. 430-52(a) and Table 430-152, the maximum rating for the largest-motor branch-circuit time-delay fuses would be not more than 175A (100A × 1.75). And, the sum of the highest rating plus the full-load currents of all other motors would also call for a fuse rated at 175A, the next lower standard rating below 180A (175A + 5A). Therefore, as ridiculous as it sounds, the maximum rating for both the largest-motor branch circuit and the feeder protective device in this case is 175A, even though we are protecting the same motors and motor circuits as before! However, in this case, the 175A-rated one-time fuses on the feeder will probably not carry the starting current of the motor and open the circuit on startup. *But,* use of higher-rated one-time fuses on the feeder would be a violation of Sec. 430-62(a) (Figure 8).

Although the basic rule does *not* generally require any consideration of the types of devices selected when determining maximum ratings for motor feeder protection, the new Exception to Sec. 430-62(a) *does* where an instantaneous-trip breaker or MSCP is used on any branch circuit supplied by the feeder to be protected. That is, where one or more instantaneous-trip CBs or MSCPs are used to provide short-circuit and ground-fault protection on the branch circuits to be supplied, then the maximum rating of the feeder protective device must be calculated with respect to the type of device selected for feeder protection.

What is the maximum permitted rating
for a time-delay fuse on the feeder?

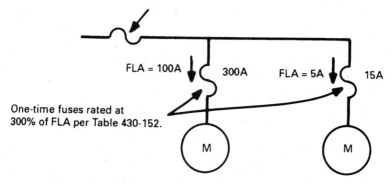

One-time fuses rated at
300% of FLA per Table 430-152.

Max rating for feeder protection = 300A + 5A
= 305A, which would result in the selectioı
of the next lower-rated 300A device. . .

. . .*but*, if the device types are reversed, one-time fuses
on the feeder and time-delay fuses on the branches, then. . .

Max rating feeder protection = 175A + 5A
= 180A, which would result in selection
of next lower-rated 175A rating.

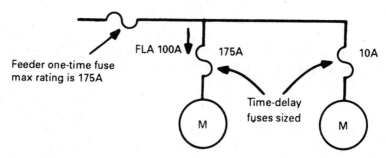

Feeder one-time fuse
max rating is 175A

Time-delay
fuses sized

Figure 8

As has been discussed, under certain conditions the Exception to Sec.
430-52(a) would permit instantaneous-trip CBs and MSCPs to be set as
high as 13 times motor full-load current. And, if the basic rule for establishing the maximum permitted rating for feeder protection were
applied, the calculated maximum value would be based on the 1300%
value of the instantaneous-trip CB or MSCP. Because that will result

in an unacceptably high rating, the new Exception to Sec. 430-62(a) now requires that the rating of the instantaneous-trip CB(s) or MSCP(s) on the branch circuit(s) be recalculated at the value permitted by Table 430-152 for the type of device that is to be used as feeder protection. Let's take an example. Consider the application shown in Figure 9.

As can be seen, there are three, 3-phase, 230V, 50hp motors protected by instantaneous-trip breakers. According to Table 430-150, the full-load current is to be taken as 130A.

Although such devices are generally limited to a trip-setting of no more than 700% of motor full-load current, here the instantaneous-trip breaker on each branch is set at 13 times full-load current—or 1690A—to prevent nuisance-tripping on startup as permitted by the Exception for instantaneous-trip breakers in Sec. 430-52(a).

The maximum permitted rating for the feeder protective device according to the 1990 and previous editions of the NEC would have been

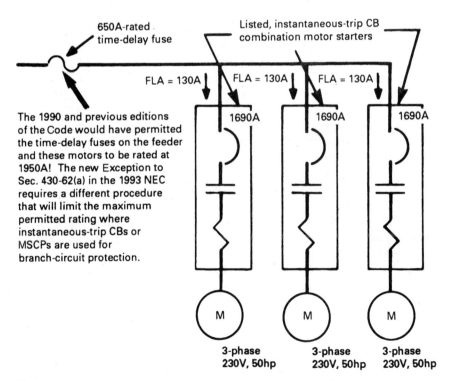

Figure 9

the sum of the highest-rated device plus the full-load currents of the other motors, or

Max Rating of Feeder Protection = 1690A + 130A + 130A

Max Rating of Feeder Protection = 1950A

As shown, the maximum rating would be 1950A, *regardless* of whether the feeder protective device is a one-time fuse, or a time-delay fuse, or an inverse-time CB.

The only real difference in the new procedure is that the instantaneous-trip breaker ratings must be correlated with the type of feeder protective device to establish the rating for largest-rated branch-circuit device. Here, one of the 1690A instantaneous-trip CBs must be rerated based on the percent value that would be permitted by Table 430-152 if the same type of device protecting the feeder were used on that branch. That is, if a one-time fuse is selected for feeder protection, the highest-rated instantaneous-trip CB branch device must be taken as being equal to the maximum value permitted for one-time fuses on a branch circuit. In this case, the highest-rated device must be taken as 390A (130A × 3 from Table 430-152 for one-time fuses). The remainder of the calculation remains as before. The full-load currents of the other motors (130A + 130A, or 260A) are added to this adjusted highest rating (390A), giving a maximum rating of 650A (260A + 390A) for one-time fuses. If time-delay fuses or CBs are used to protect the feeder, the highest-rated device would be taken as equal to 175% or 250%, respectively, of the motor full-load current and summed with the full-load currents of the remaining motors.

What would be the maximum permitted rating of time-delay fuses for a feeder supplying two, 3-phase, 230V, 50hp motors where one motor was protected by an instantaneous-trip breaker and the other by a one-time fuse?

As shown in Figure 10, the instantaneous-trip breaker is set at 1690A in accordance with the Exception to Sec. 430-52(a) and the one-time fuse is rated at 350A as per Table 430-152 and Sec. 430-52(a). The adjusted rating of the instantaneous-trip breaker where time-delay fuses are to be used as feeder protection must be taken as being equal to the maximum permitted rating if time-delay fuses were used to protect that branch circuit. In this case, Sec. 430-52(a) and Table 430-152 would recognize a maximum rating of 227.5A (130A × 1.75). *But,* because the table permitted 390A rating of the one-time fuse to be greater than 227.5A, the 390A one-time fuses become the highest-rated device. And, the full-load current of the motor protected by the instantaneous-trip

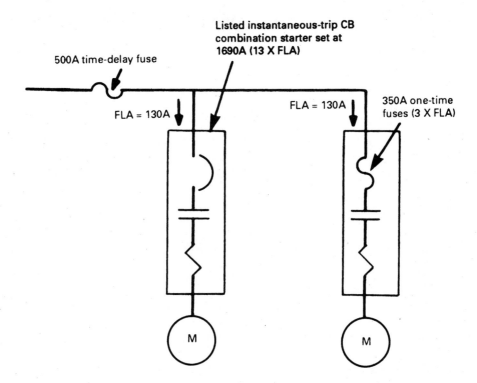

Listed instantaneous-trip CB
combination starter set at
1690A (13 X FLA)

500A time-delay fuse

FLA = 130A

FLA = 130A

350A one-time
fuses (3 X FLA)

M   M

When the rule of Sec. 430-62(a), Ex. is applied, the "adjusted rating" of
instantaneous-trip CB becomes 227.5A. Because that "rating" is lower than the
390A maximum permitted rating for the one-time fuse on the other motor, the
one-time fuses become the "largest-rated" device and the full-load current of the
instantaneous-trip CB protected motor is added to the 390A.

Max rating for feeder time-delay fuses = 390A + 130A
= 520A, or 500A

Figure 10

breaker is simply added to the ampere rating of the highest-rated
branch-circuit device. Therefore, the maximum permitted rating for
time-delay fuses on the feeder would be 520A (390A + 130A). And 500A
time-delay fuses—the required next lower rating—would be selected.

What this rule boils down to is this: instantaneous-trip breakers and
MSCPs are never to be taken at their actual settings when establishing the maximum permitted rating of feeder protection. The rating of
any instantaneous-trip CB or MSCP must be taken as being equal to
the maximum rating permitted on the given branch circuit by Sec. 430-

52 and Table 430-152 for the specific type of protective device to be used on the feeder. And, if the adjusted rating is *not* the highest when compared to the rating of any one-time fuses, time-delay fuses, or CBs used on other branch circuits supplied by the feeder, then the full-load current(s) of the motor(s) protected by an instantaneous-trip breaker or MSCP are simply added to the other full-load currents and summed with the ampere value of the highest-rated fuse or CB.

Watch Out for...
"Service-type" Bonding Requirements in Hazardous (Classified) Locations

Sec. 501-16(a) and Exception.   Examination of new exception to Secs. 501-, 502-, and 503-16(a) reveals a not-so-obvious inconsistency—and potential inadequacy—where "service-type" bonding of raceways and enclosures is provided to assure low-impedance fault-return path in hazardous location installations.

As most are aware, installations of electrical systems and equipment in so-called "Hazardous (Classified) Locations" are treated in a different manner than installations in other, nonhazardous, locations. In recognition of the additional concerns associated with safely supplying and using electrical power in areas where the processes produce explosive concentrations of combustible vapors, liquids, dust, fibers, etc., the NEC logically requires the use of different equipment and techniques to minimize the possibility of accidental ignition of the combustible material.

Of the many NEC requirements for the use of specialized equipment and/or techniques, one of the more important requirements is the rule that covers bonding of raceways and enclosures within Class I, II, and III Hazardous (Classified) Locations. Special care in the grounding of all equipment is necessary in Div 1 and Div 2 locations to minimize the possibility of arcs or sparks during fault conditions at the points where any grounded metal is interconnected. To prevent such an occurrence, Secs. 501-16(a), 502-16(a), and 503-16(a)—which cover Class I, II, and III Locations, respectively—essentially require that all connections of conduit to boxes, cabinets, enclosures for apparatus, and motor frames must be made using one of the "service-type" bonding methods. And that requirement is repeated in Sec. 250-78, which specifies the use of the bonding techniques recognized for services by "Sec. 250-72(b) through (e)." (See Figure 1.)

The rules covering bonding within Hazardous (Classified) Locations have been amended in each new Code since the 1984 edition. And in each edition the change in wording has been aimed at clarifying where the required bonding techniques must be applied. The new Exception to Secs. 501-, 502-, and 503-16(a) in the 1993 NEC also qualifies the extent to which the service-type bonding must be used. *But,* there is an apparent lack of consistency in treatment of similar applications.

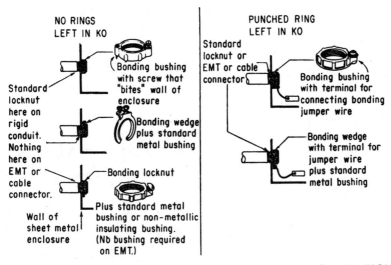

Figure 1 The bonding methods required for services given in Sec. 250-72(b) through (e). The same methods are also required for bonding of all raceways and enclosures in Div 1 and Div 2, Class I, II, and III Hazardous (Classified) Locations.

Evolution of the Rules

In the 1984 NEC, the rules for bonding of raceways and enclosures within hazardous areas only literally required service-type bonding for "wiring and equipment *in*" the classified locations. However, to be truly effective, this method of installation is not only necessary within hazardous areas, but also needs to be carried back to the point where the electrical supply is bonded and grounded to assure quick fault-clearing action of the circuit protective device. For example, if the service-type bonding were only used within the classified spaces, a high-impedance connection outside the hazardous area could prevent adequate current flow back to the system neutral to cause operation of the fuse or breaker protecting the faulted circuit. A voltage on the non-current-carrying metallic parts within the hazardous location represents both a shock *and* an explosion hazard because it increases the possibility of an open spark or arc and that is exactly the type of dangerous condition that the required service-type bonding is intended to prevent (Figure 2).

That shortcoming was recognized and the wording of the rules for bonding within hazardous areas was amended in the 1987 edition. In recognition of the need for greater reliability in the ground return path all the way back to the point of system bonding and grounding and

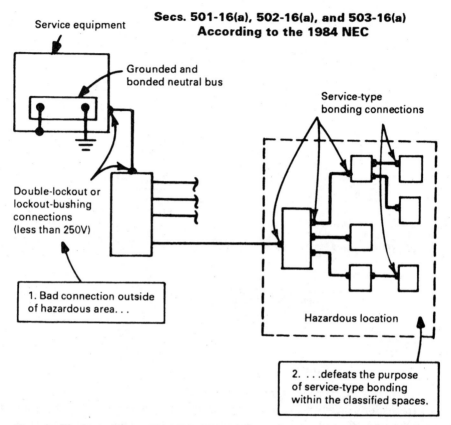

Secs. 501-16(a), 502-16(a), and 503-16(a)
According to the 1984 NEC

Service equipment

Grounded and
bonded neutral bus

Service-type
bonding connections

Double-lockout or
lockout-bushing
connections
(less than 250V)

1. Bad connection outside
of hazardous area. . .

Hazardous location

2. . . .defeats the purpose
of service-type bonding
within the classified spaces.

Figure 2 Wording of Secs. 501-16(a), 502-16(a), and 503-16(a) in the 1984 NEC only required service-type bonding connections within the hazardous spaces. But, to be truly effective, such connections should really be carried back to the "point of grounding." This is especially true of grounded systems where the service-type bonding is now required from the classified area all the way back to where the neutral is bonded and grounded.

because such bonding and grounding is generally provided at the service equipment, the 1987 Code specifically required the service-type bonding methods to be provided throughout the hazardous area and all the way back to "the point of grounding of the service equipment."

While that wording adequately addressed the need for greater ground-return path reliability back to the neutral-ground connection-point, it did not consider the fact that a hazardous location may be supplied from a transformer located between the service equipment and the hazardous area. And the literal wording of the rules as they appeared in the 1987 Code required the service-type bonding all the way back to the service. But, because Sec. 250-26 generally requires

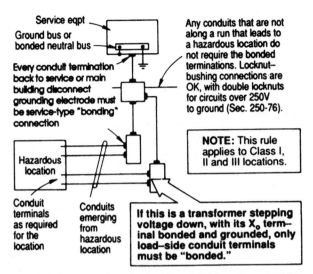

Figure 3 Example of the basic rule, which was changed in the 1990 NEC to require service-type bonding back to the "point of grounding for service equipment or the point of grounding of a separately derived system." This remains the basic rule of Secs. 501-16(a), 502-16(a), and 503-16(a) in the 1993 NEC.

bonding and grounding of such a "separately derived system" as if it were a service, there is really no need to go beyond that point. That is, because the neutral (grounded) conductor is bonded to the ground bus and the grounding electrode system (nearby building steel) on the transformer's secondary side, the service-type bonding method need only be carried back to the enclosure where those connections are made. And that is all that is necessary to assure the required and desired ultra-high-reliability return path to the "point of grounding." As a result, a change was made to the wording of Secs. 501-, 502-, and 503-16(a) in the 1990 NEC to indicate that the service-type bonding need not be carried back beyond a separately derived system to the service (Figure 3). The basic rule of those sections remains unchanged in the 1993 Code.

Defining the Problem

As we have seen, the rules governing application of the service-type bonding methods with raceways and enclosures that supply or occupy hazardous spaces have been changed over the past several editions of the Code to clarify that the specified bonding must be provided from the

The basic rule of Sec. 250-24(a)
for grounded systems.

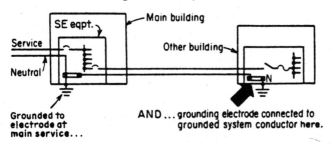

The basic rule of Sec. 250-24(b)
for ungrounded systems.

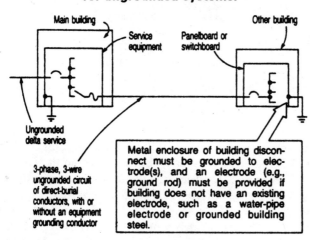

Figure 4 The basic rule of Sec. 250-24(a) and (b). Only these two additional applications were included in the original proposal to add the new Exception to Secs. 501-16(a), 502-16(a), and 503-16(a) in the 1993 NEC. Because the main building disconnects are bonded and grounded as if they were service equipment, there is no need to continue the service-type bonding methods beyond the main building disconnects.

bonded and grounded neutral bus—or ground bus in an ungrounded system—all the way through to the last outlet box in the classified area. For the 1993 NEC, a very well-considered proposal was submitted to recognize that, under certain conditions, there is no need to carry the service-type bonding of raceways and enclosures beyond the main building disconnects. That is, where one building is supplied from a service or separately derived system in another building, the basic rules of Sec. 250-24(a) and (b) require that the outbuilding main disconnect be grounded and bonded as if it were a service. That is, for grounded systems, Sec. 250-24(a) requires bonding of the neutral bus to the equipment ground bus and the outbuilding grounding electrode system at the main building disconnect—just as if it were a service disconnect. And the basic rule of Sec. 250-24(b), for ungrounded systems, also requires bonding of the equipment ground bus to the outbuilding grounding electrode system at the main building disconnect. It was pointed out that because those main building disconnects are bonded and grounded as if they were a service, there is really no need for continuing the service-type bonding of raceways and enclosures beyond those main building disconnects to the service or separately derived system. The Code Making Panel agreed and added an Exception to the basic requirement for main building disconnects that are bonded and grounded in accordance with Sec. 250-24(a) and (b) (Figure 4).

All is well and good up to this point. However, during the Comment period of the Code Making Cycle, a comment was submitted to include those main building disconnects grounded and bonded in accordance with Sec. 250-24(c) in the newly accepted Exception because:

> Sec. 250-24(c) also establishes rules for building disconnecting means that are remote from the building served, and must be correlated here.

While it is true that Sec. 250-24(c) *does* establish rules for main building disconnects, the grounding at the outbuilding supplied by the referenced remote disconnects is *not* completed as if it were a service!

The rule of Sec. 250-24(c), which was new in the 1990 NEC, is essentially the same as Exception No. 1 to Sec. 250-24(a). That is, Sec. 250-24(c) addresses those multibuilding industrial installations where outbuildings are supplied from a service in another building *and* the disconnect is *not* located in the outbuilding, as covered in Sec. 225-8 (b), Ex. Nos. 1 and 2. Where the disconnect is located remote from the outbuilding as permitted by the exceptions to Sec. 225-8(b), the rule of Sec. 250-24(c) expressly *prohibits* connection of the neutral bus to the ground bus (Figure 5).

Although Sec. 250-24(c) also requires connection of the ground bus to a grounding electrode system—as would be required at a service or separately derived system—as far as fault-clearing is concerned, connection to earth through the grounding electrode system is not critical. It is the neutral-ground connection that facilitates current flow to cause blowing or tripping of the circuit protective device. And such connections are generally only to be made at a service, a separately derived system, or main building disconnect grounded in accordance with Sec. 250-24(a), which is why service-type bonding methods need only be car-

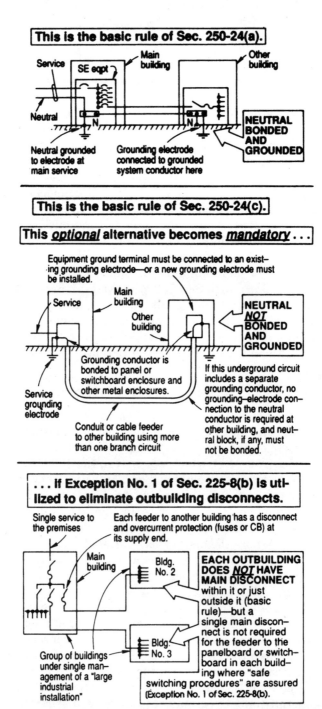

This is the basic rule of Sec. 250-24(a).

Service — SE eqpt — Main building — Other building

Neutral

NEUTRAL BONDED AND GROUNDED

N — N

Neutral grounded to electrode at main service

Grounding electrode connected to grounded system conductor here

This is the basic rule of Sec. 250-24(c).

This *optional* alternative becomes *mandatory* . . .

Equipment ground terminal must be connected to an existing grounding electrode—or a new grounding electrode must be installed.

Service — Main building — Other building

NEUTRAL *NOT* BONDED AND GROUNDED

Service grounding electrode

Grounding conductor is bonded to panel or switchboard enclosure and other metal enclosures.

Conduit or cable feeder to other building using more than one branch circuit

If this underground circuit includes a separate grounding conductor, no grounding–electrode connection to the neutral conductor is required at other building, and neutral block, if any, must not be bonded.

. . . If Exception No. 1 of Sec. 225-8(b) is utilized to eliminate outbuilding disconnects.

Single service to the premises

Each feeder to another building has a disconnect and overcurrent protection (fuses or CB) at its supply end.

Main building

Bldg. No. 2

Bldg. No. 3

Group of buildings under single management of a "large industrial installation"

EACH OUTBUILDING DOES *NOT* HAVE MAIN DISCONNECT within it or just outside it (basic rule)—but a single main disconnect is not required for the feeder to the panelboard or switchboard in each building where "safe switching procedures" are assured (Exception No. 1 of Sec. 225-8(b).

Figure 5 The rule of Sec. 250-24(c) does not provide for the same bonding and grounding that is required at a service, separately derived system, or main building disconnect installed in accordance with the basic rule of Sec. 250-24(a). Therefore, where the hazardous area is located in an outbuilding that is grounded and bonded in accordance with Sec. 250-24(c), serious consideration should be given to continuing the service-type bonding methods for all raceways and enclosures back to the service or separately derived system.

ried to any one of those points in the system. *But,* because there is no neutral-ground connection at the outbuilding when grounded and bonded as *required* by Sec. 250-24(c), the service-type bonding methods *should not* stop there. The bonding of raceways and enclosures called for by the basic rule *should be* continued back to the first location where the system neutral is grounded and bonded, i.e., service or separately derived system.

It seems as if the Comment that suggested inclusion of Sec. 250-24(c) in the new Exception to the bonding requirements for hazardous areas, as well as the Panel Action to accept that suggestion, was not clearly thought out. Without a doubt, such application would be a return to the objectionable condition that was permitted under the 1984 and previous Codes, and, has since been prohibited. That is, the service-type bonding of raceways and enclosures would not be continued as far as necessary to be completely effective. The original proposal correctly identified those two additional applications that would be required by the previous wording to carry the service-type bonding beyond that which is necessary.

Another problem with any attempt to use this Exception relates to the location of the "point of grounding of a building disconnecting means." For main building disconnects installed in accordance with Sec. 250-24(a) or (b), the "point of grounding" is at the disconnect. The "point of grounding" for a remote disconnect supplying an outbuilding is very difficult to establish. For services, separately derived systems, and certain main building disconnects, the "point of grounding" is that point where the system is bonded and grounded. In reality, the "point of grounding" for the remote disconnect is the service or separately derived system supplying the outbuilding disconnect. That becomes especially apparent when one considers that outbuildings with remote disconnects are essentially treated as a subpanel within the same building (i.e., neutral-ground connection is *prohibited*), and they should be viewed as such. That is, it would not be permissible to discontinue the service-type bonding at a subpanel within the same building, so it should not be permissible to discontinue those methods at the outbuilding, or even at the remote disconnect.

Obviously, the way that anyone handles this apparent inconsistency will be up to that individual. But, there are a few things to be considered:

- inclusion of Sec. 250-24(c) was recommended and accepted during the comment period, and not subject to the amount of scrutiny and consideration afforded to a full-fledged Proposal
- the functional inadequacy that would result from actual application

■ wording ("point of grounding of a building disconnecting means") of the new Exception makes identifying that location difficult when applied to a remote main building disconnect

In light of the preceding revelations, it would seem most prudent to ignore the permission given by the new Exception for remote main building disconnects. Remember, compliance with the Exception is not *mandatory*, it is *permissible*. Opt not to exercise that permission. Provide the high-reliability bonding methods all the way back to the service or separately derived system where an outbuilding is supplied by a remote disconnect and installed as specified in Sec. 250-24(c).

Nomographs
Show IC Ratings
of Fuses and CBs

Sec. 110-9. Given the available rms fault current at a service or transformer output, just use a straight edge to determine short-circuit fault current at the end of any given length of cables from No. 8 up to 750 kcmil copper, 277 and 120VAC, in either steel or aluminum conduit. Curves are also given for determining short-circuit current on the secondary side of 480-208Y/120VAC transformers.

The nomographs here provide a handy and easily used technique for everyone's use in assuring compliance with the strict and very important rule of **NEC** Sec. 110-9, which says:

110-9. Interrupting Rating. Equipment intended to break current at fault levels shall have an interrupting rating sufficient for the system voltage and the current which is available at the line terminals of the equipment.

That **Code** rule requires that each and every fuse and circuit breaker must have a short-circuit interrupting rating in symmetrical rms amperes that is at least equal to the fault current the circuit can deliver into a bolted phase-to-phase short on the ungrounded conductors of a circuit. The nomographs indicate the minimum short-circuit duty rating required of a protective device for the available current at the line (not load) terminals of the fuse or CB.

Figure 1

In this age of computers, many people might wonder about the value of nomographs for determining reduction of 3-phase short-circuit currents in cables and in 480-208Y/120VAC transformers. Certainly nomographs should never be viewed as an ideal substitute or replacement for basic short-circuit calculations, as the engineering profession has always made them.

But for the very many instances when a complete set of electrical drawings for a given project have to be checked fairly quickly in a limited time frame, nomographs provide an extremely valuable tool for engineers, contractors, electricians, and inspectors. When you're confronted with checking schedules on panelboards, motor controllers and control centers, and switchboard/switchgear schedules, and you find

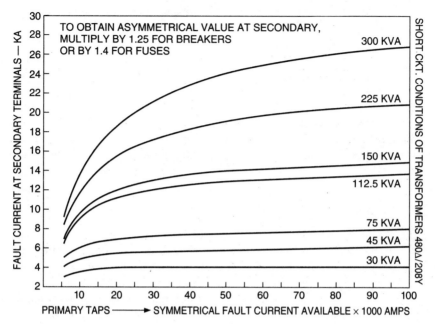

Figure 2

out that short-circuit ratings are not shown (and very often were never calculated), then the nomographs shown in this series can help to reduce the panic and provide a reading on the safety characteristics of the overall systems.

Just how good are these nomographs? In making short-circuit calculations, the first step is to construct a one-line diagram of the system. From the one-line diagram, the impedances of sources (service, generators, motors), feeders, transformers, etc., are then developed. From this data an impedance diagram is constructed. Then the impedances are added vectorially, then reduced and combined until there is only one impedance.

When the voltage of the source is divided by the impedance that has been determined—following Ohm's Law—the three-phase fault current is calculated. Of course, this is all a relatively simple explanation of a procedure that can be rather complex and involved. But the explanation given here is intended to make a point: The cable short-circuit diagrams were developed from an equation reading as follows:

$$1/A = 1/B + 1/C$$

Although that would represent an arithmetic addition of imped-ances, when the calculations were made, they were actually made vec-torially. As a result, when these nomographs are used, the actual value of short-circuit current indicated on the nomograph will be correct or slightly higher than the real calculated values. Thus, the short-circuit values shown will be conservative and on the safe side.

When plotting the reduction of short-circuit currents through trans-formers, the values of impedance used were the lowest manufacturers' impedances that could be found. So using these nomographs will again yield results that are always on the safe, conservative side of actual values.

Let's Take an Example

Take a look at the nomograph for 500kcmil cables at 277 volts (page 122). Assume this cable terminates at a panelboard, and the fault cur-rent rating of the panelboard is to be determined (Figure 3).

Assume a fault current of 150,000 rms amperes is available at the switchboard that feeds the panel. Say the length of the 500kcmil feeder run is 150 ft. Put one end of your straight edge at 150 on the left verti-cal scale of available fault current. Put the other end of the straight edge at 150 feet on the right vertical scale of cable length in feet. Then read the number where the straight edge crosses the two center scales which show the available fault current at the end of the 150-ft. feeder run.

By interpolation, you will read:

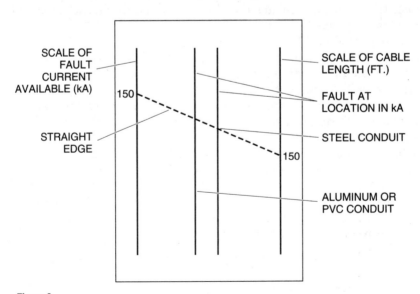

Figure 3

32,000A for the circuit in aluminum conduit

or

29,000A for the circuit if it is in steel conduit.

Conclusion: The panelboard and CBs or fuses at the load end of the 150-ft. feeder run must have a short-circuit rating of *either* 29,000A or 32,000A, depending on the material of the conduit or raceway enclosing the circuit.

Transformer Plot

For using the transformer curve shown in Figure 2, the procedure is even simpler.

Take a 75kVA transformer with a primary available fault current of 80,000 rms amps. Run a line up vertically from the value of 80,000 on the horizontal coordinate at bottom (Primary-Side fault current available) to its intersection with the curve for the 75kVA transformer. Then from that intersection point, run a line horizontally to the left. Note that the horizontal line intersects the left side coordinate (Fault Current on transformers secondary terminals) at 7.6 kA, or 7600A on the 208Y/120VAC secondary side. Note that above 100,000A, the curves are nearly flat. Therefore, the 100,000A figure can be used for all fault currents exceeding 100,000A on the primary side of the transformer (Figure 4).

As can be seen, these nomographs do not solve all problems. But for low-voltage projects, they can be a great help.

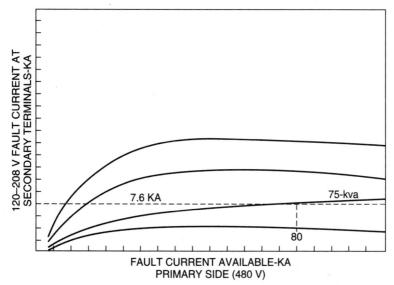

Figure 4

#8 Cu. CABLE – 120 VOLTS

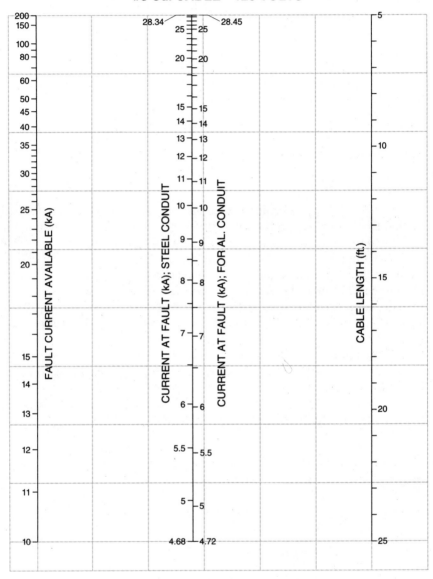

#6 Cu. CABLE – 120 VOLTS

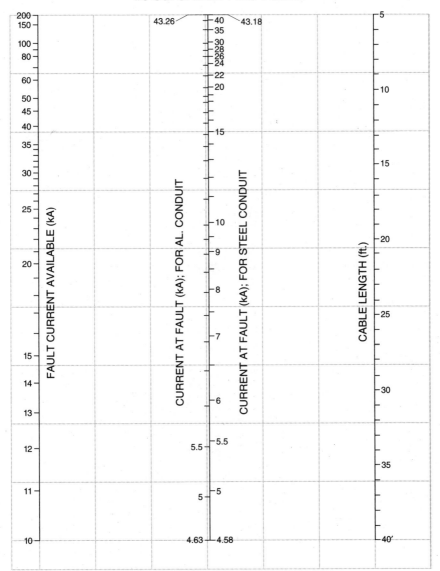

#4 Cu. CABLE – 120 VOLTS

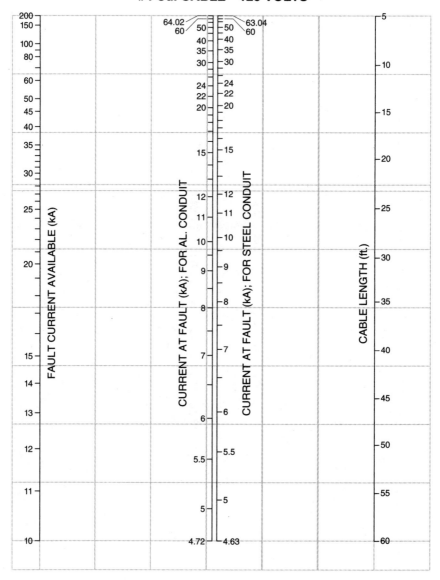

#2 Cu. CABLE – 120 VOLTS

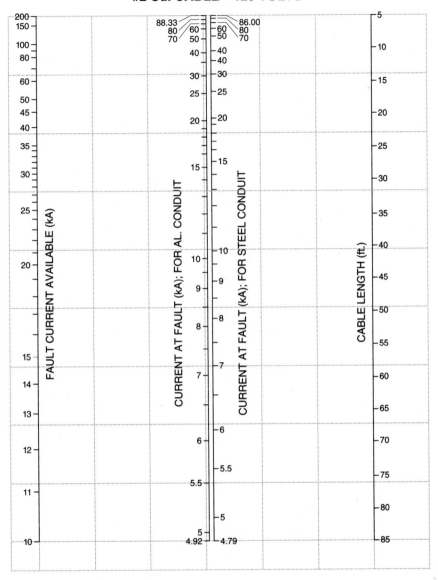

#1 Cu. CABLE – 120 VOLTS

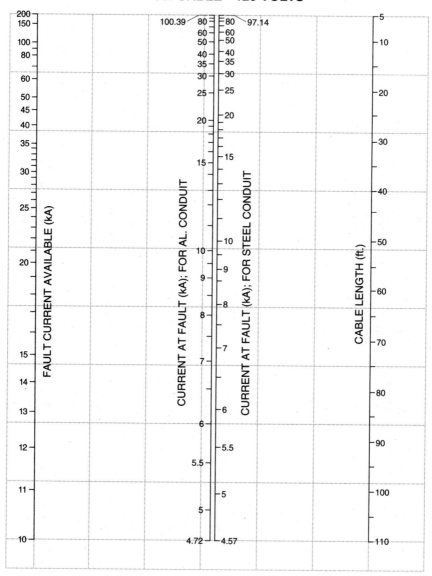

#1/0 Cu. CABLE – 120 VOLTS

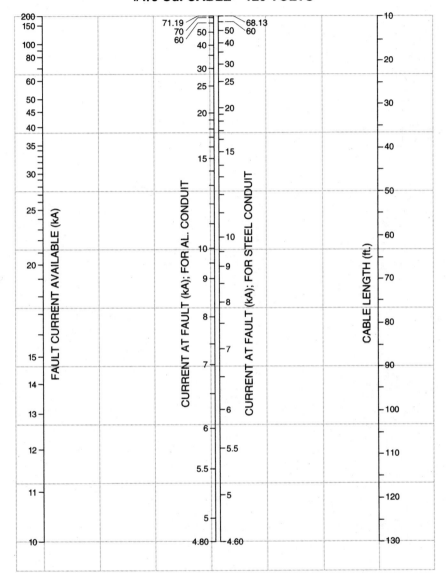

#2/0 Cu. CABLE – 120 VOLTS

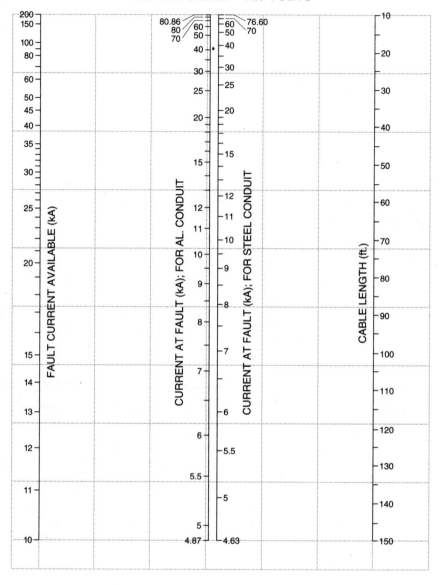

#3/0 Cu. CABLE – 120 VOLTS

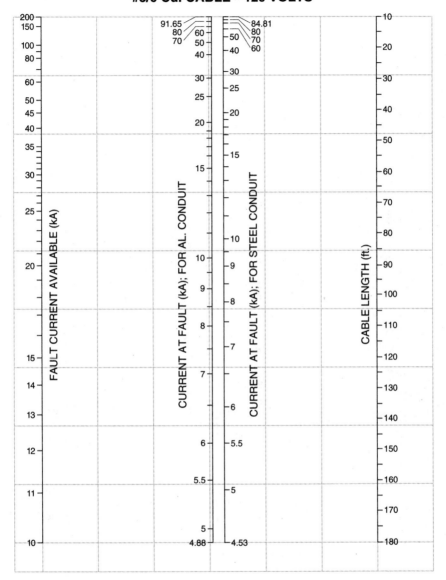

#4/0 Cu. CABLE – 120 VOLTS

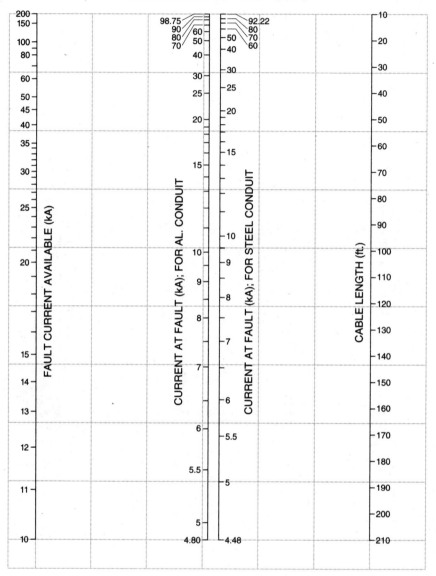

250 kcmil Cu. CABLE – 120 VOLTS

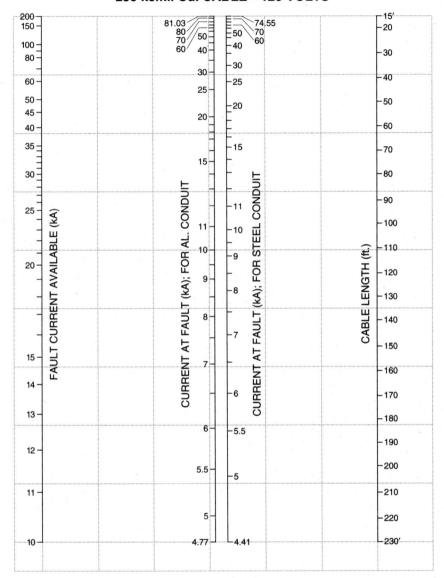

300 kcmil Cu. CABLE – 120 VOLTS

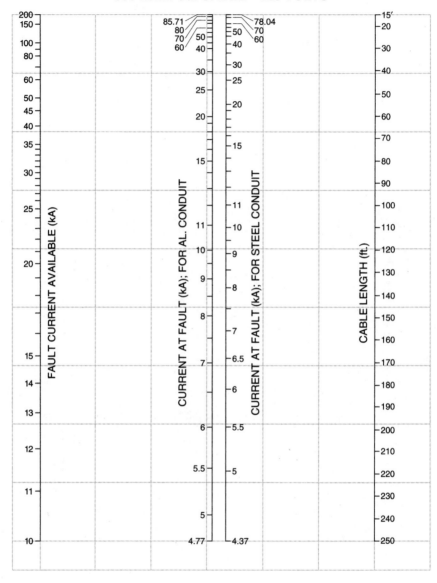

350 kcmil Cu. CABLE – 120 VOLTS

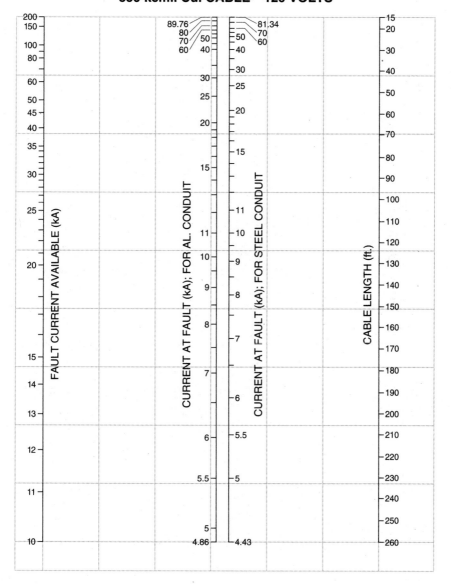

400 kcmil Cu. CABLE – 120 VOLTS

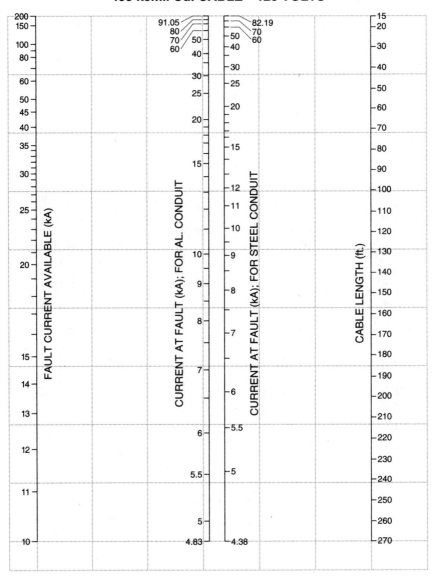

500 kcmil Cu. CABLE – 120 VOLTS

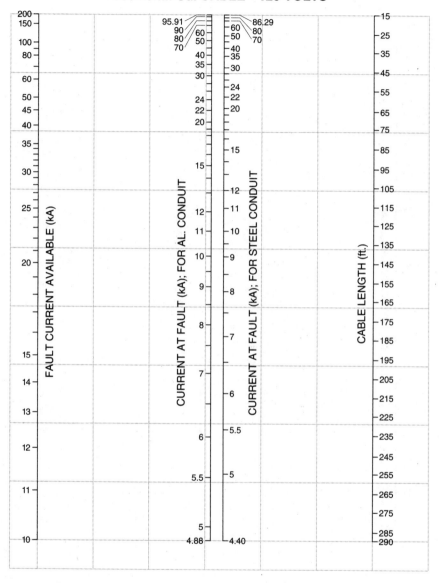

600 kcmil Cu. CABLE – 120 VOLTS

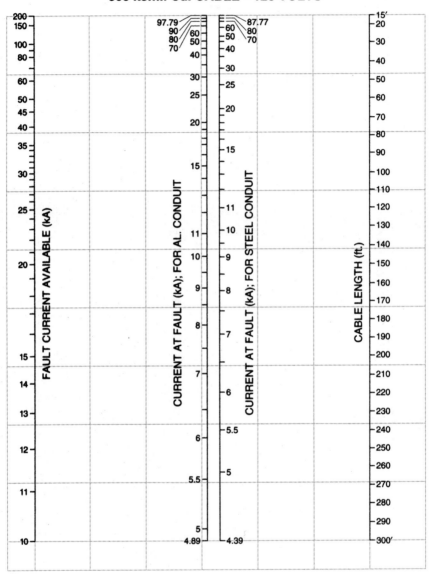

750 kcmil Cu. CABLE – 120 VOLTS

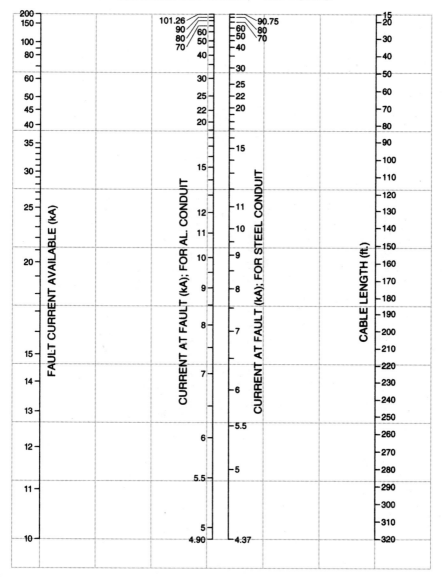

#8 Cu. CABLE – 277 VOLTS

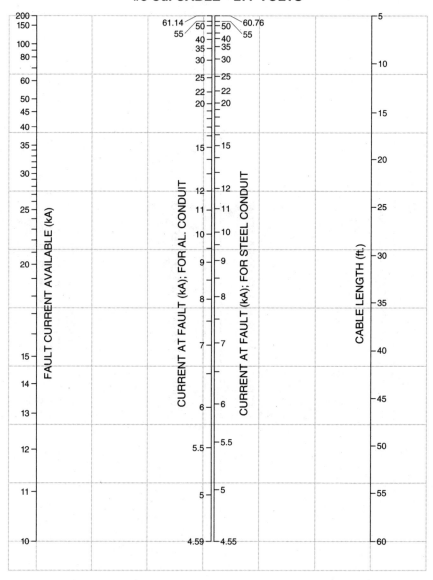

#6 Cu. CABLE – 277 VOLTS

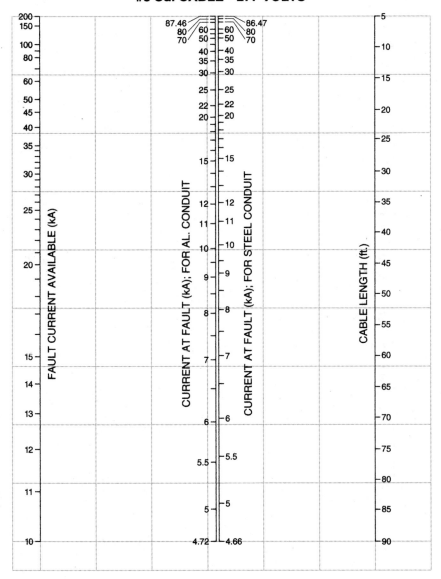

#3 CU. CABLE – 277 VOLTS

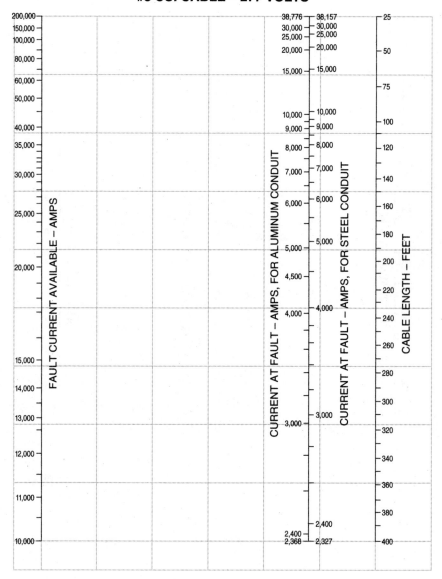

#4 CU. CABLE – 277 VOLTS

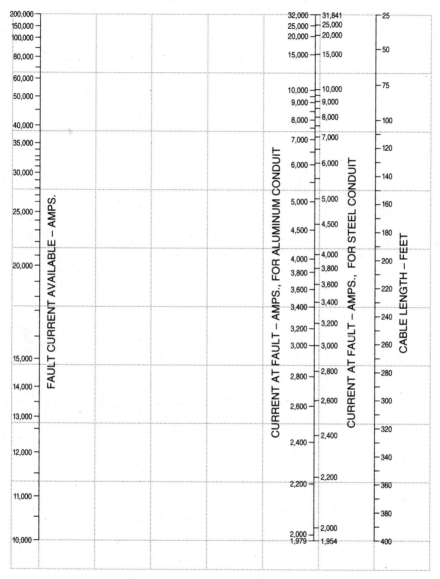

#2 CU. CABLE – 277 VOLTS

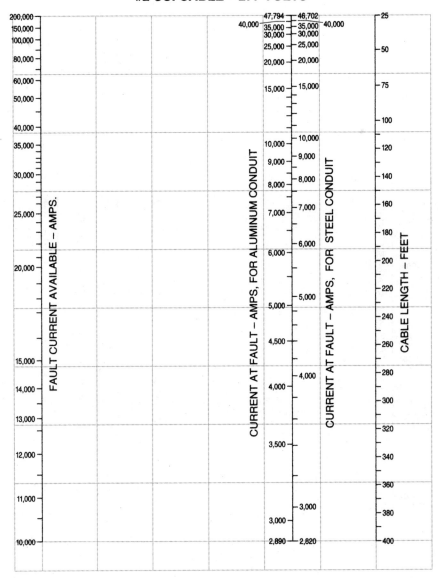

#1 Cu. CABLE – 277 VOLTS

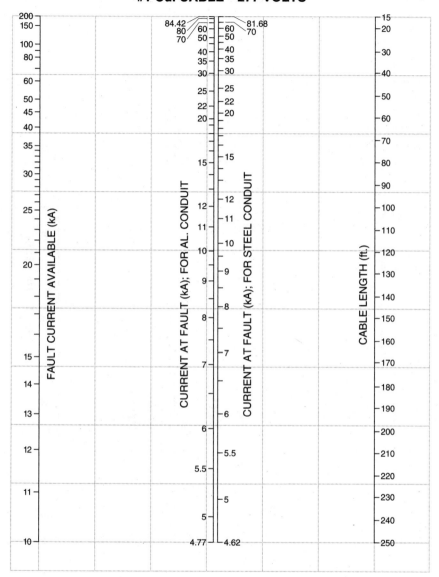

#1/0 CU. CABLE – 277 VOLTS

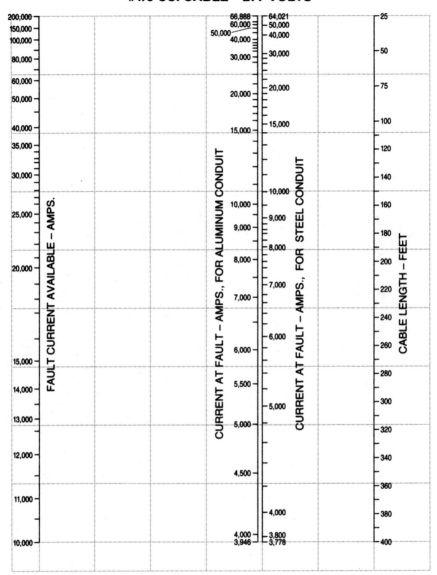

#2/0 Cu. CABLE – 277 VOLTS

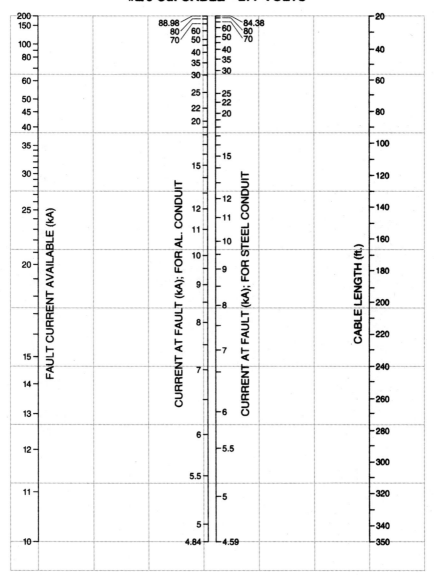

3/0 COPPER CABLE – 277 V

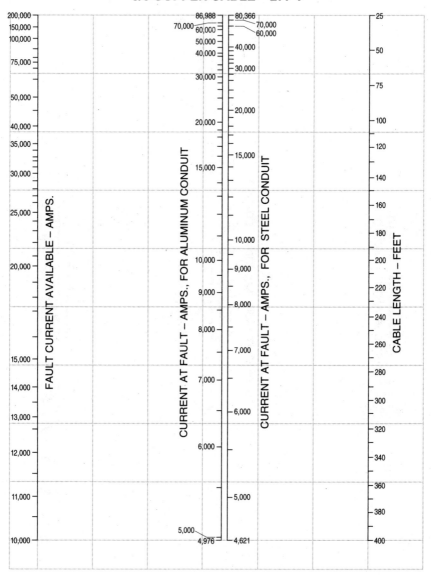

4/0 COPPER CABLE – 277 V

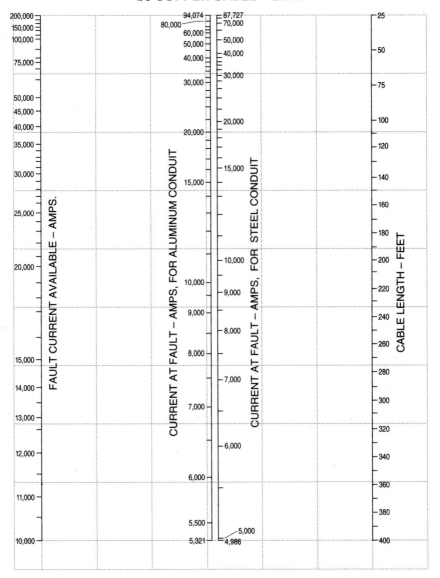

250 kcmil CU. CABLE – 277 V

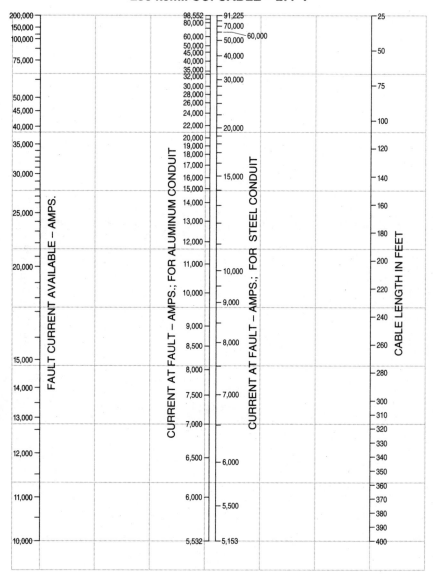

300 kcmil Cu. CABLE – 277 VOLTS

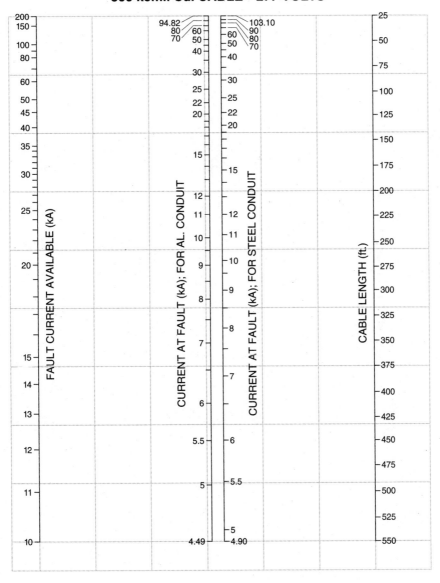

350 kcmil CU. CABLE – 277 V

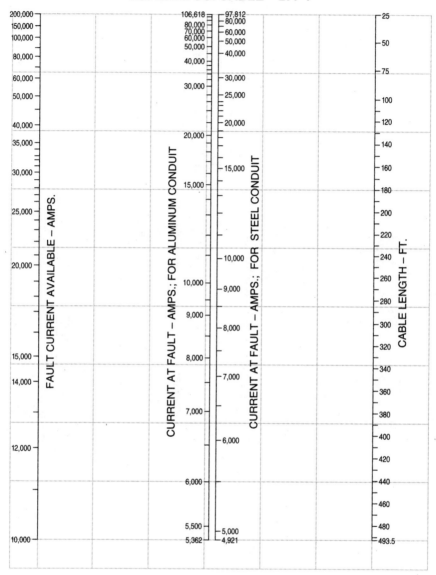

400 kcmil Cu. CABLE – 277 VOLTS

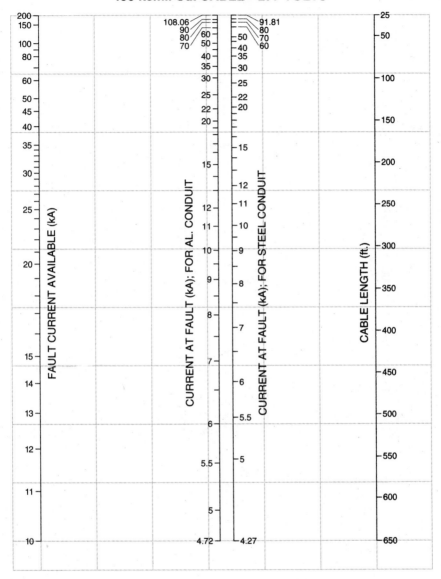

500 kcmil CU. CABLE – 277 V

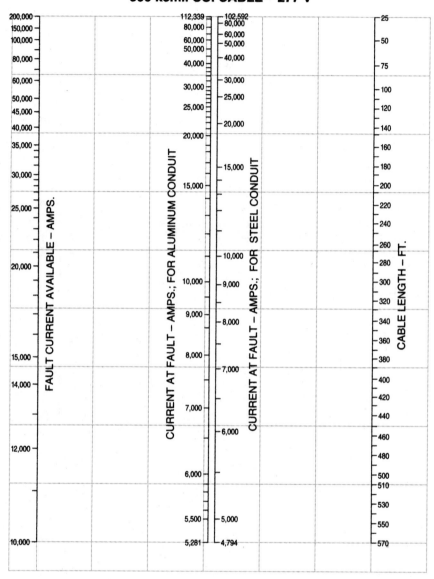

600 kcmil Cu. CABLE – 277 VOLTS

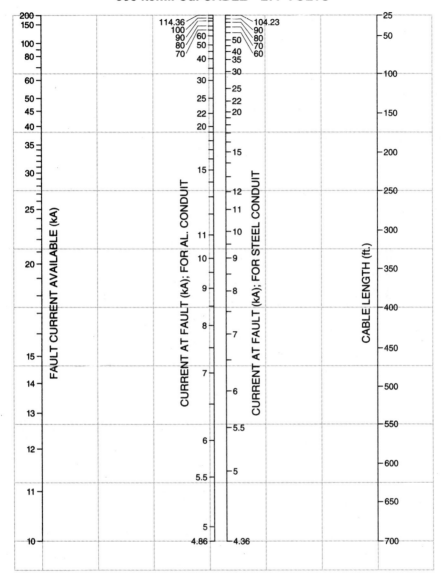

750 kcmil Cu. CABLE – 277 VOLTS

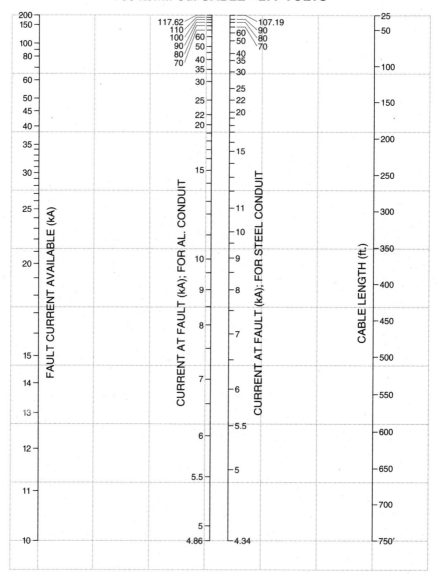

Index